Letters of a Ticonderoga Farmer

A $15,000 LAWSUIT!

ABOUT A FOUR YEAR OLD COLT.

[☞ Re-printed from the Whitehall CHRONICLE of January 24th, 1874.—EDITOR.]

TICONDEROGA, January 20, 1874.

Editor of the Whitehall Chronicle.

☞ THE COOK HORSE CASE.—In your issue of the 17th inst., you describe the proceedings of a day at Judge Potter's Chambers in your village, among which appears the case of Cook vs. Bailey, upon a motion by plaintiff to restrain the defendant from trotting the Ticonderoga colt "Kitty Cook" in races for wagers or otherwise, during the pendency of the suit. A brief outline may be interesting, and is as follows:

In the spring of 1873 Wm. H. Cook, of our town, gave permission to his friend Dr. J. H. Bailey, (a practicing physician) to drive and use his colt "Kitty Cook," then three years old, past. The doctor claims that soon after this he made a bargain for the purchase of the colt at $700, with a long time for payment. Mr. Cook's version is that he offered the colt for $700, and there was talk about a trade, but no such condition as to time of payment was made, as claimed by the doctor. In a few weeks Mr. Cook called upon the doctor and told him if he considered their former talk a trade, he must pay him the $700 then. The doctor replied that he was not then in readiness to pay, but would secure Mr. Cook if he had any fears. Mr. Cook declined security, and told the doctor if he kept the colt, he must comply with the same conditions that W. G. Baldwin took the colt upon the fall before. The bargain with Mr. Baldwin was, that he should develope the speed of the colt, and pay Mr. Cook $700 when sold, together with one half of the amount he should obtain over that sum. The doctor denies assenting to this. During all this, as time flew on, the colt flew on—daily increasing in speed until the "forties" were of no account. Still on, in a twinkling, as it were, the "thirties" were left in the shade, and the colt fairly revelled in the "twenties," by showing half miles in 1:11 and 1:12. The excitement became intense and the horse fever raged so high that it was declared by visitors from across the lake and by one Sanford, in particular, that our town was a land of lunatics, having but one sane man in it—he being too aged to *talk horse*.

Doctor Bailey tendered Mr. Cook $700, which was refused, and a suit was brought in replevin for the possession of the mare, claiming her to be worth $3,000. The doctor gave the requisite bond to the sheriff and resumed the possession. Mr. Cook now makes the motion you speak of, asking for an injunction to restrain the doctor from trotting the colt in races for wagers or otherwise, during the pendency of the suit, showing, by affidavits, as you say, that the colt has shown a speed of 2:24, and to be worth at least, $15,000. Upon this motion you say Judge Potter has reserved his decision. The case will be brought to trial at the Essex Circuit, early in June. In the mean time let the public suspend judgment upon this *flighty* dispute between two old neighbors and former friends. ABRAHAM.

[Since receiving the above letter a decision has been rendered by Judge Potter, which leaves doctor Bailey free from any restraint in the management of the colt "Kitty Cook."—EDITOR.]

The colt, "Kitty Cook," and the Boston colt, "Ticonderoga," (formerly Bay Azro) are from Messrs. Bates & Baldwin's horse, "ABRAHAM," [by Daniel Lambert, by Ethan Allen, by Black Hawk,] at the Cream Hill Stock Farm, Shoreham, Vt.

PRO BONO PUBLICO.

Letters of a Ticonderoga Farmer

SELECTIONS FROM THE CORRESPONDENCE OF WILLIAM H. COOK AND HIS WIFE WITH THEIR SON, JOSEPH COOK, 1851–1885, AS EDITED BY

Frederick G. Bascom

FALL CREEK BOOKS
AN IMPRINT OF
CORNELL UNIVERSITY PRESS
ITHACA AND LONDON

First published 1946 by Cornell University Press
First printing, Fall Creek Books, 2009
Printed in the United States of America

Contents

Introduction

SOUTH of the famous village of Ticonderoga in Essex County, New York, runs Trout Brook Valley. Through the valley passes the highway that connects Lake George and Lake Champlain. On the west side of this highway and about two miles outside of Ticonderoga village there are to be seen the tumble-down remnants of what was once a house of some pretension. In the days of its prosperity it was called Cliff Seat and was the home of Joseph Cook, who in the last two decades of the last century was a man of world-wide fame. Today he belongs among those whom someone has called the illustrious obscure. No one has written his biography, even in this era when the writing of biographies is a popular business. He is forgotten today and was almost forgotten before he died. There is an excellent sketch of him in the *Dictionary of American Biography,* the author of which tells us that Joseph Cook's death in 1901 "did not attract widespread attention." His career furnished a good example of the ancient words—*sic transit gloria mundi.*

Cliff Seat was also the home of Joseph Cook's parents, Mr. and Mrs. William H. Cook. Although Joseph Cook did not attain a place among the immortals, the renown that he enjoyed while he lived fulfilled the early ambition that William H. Cook formed for his son. In pursuance of his resolve to give the boy the best education that it was possible to have, he began sending him away to school in the youngster's thirteenth year. After that Joseph never returned beneath the parental roof except for short periods.

But father and son kept in close touch by correspondence, with Mrs. Cook frequently adding postscripts to her husband's letters. The correspondence was preserved on both sides from 1851 when it began until 1885 when William H. Cook died.

A few years ago these letters came into my possession. From them I have selected those which appear in the following pages. Aside from the local historical interest that they bear, they are filled with human interest and present an excellent picture of the manner of life followed by the rural inhabitants of northern New York during the period covered by the correspondence.

Trout Brook Valley was an inviting haven to those pioneers who came from New England after the Revolutionary War, approaching either through Vermont or up the valley of the Hudson. There was timber for every purpose, an abundant supply of water, and rich grazing for flocks and herds while the brook near at hand yielded fish in plenty, and even from the door of his cabin the rifleman could bring down venison for his table. According to an early account, it was reported among the pioneers as they made their way thither that at Ticonderoga were to be found pigs and fowls ready cooked, running around with knives and forks stuck in their backs and crying, "Eat us!"

Among the ambitious men and women who left Massachusetts and Connecticut and drove their ox teams through the forests into Vermont and northern New York in search of new settlements was Samuel Cook. He arrived at Ticonderoga in 1796 and made a clearing near the spot where Abercrombie's men were torn to pieces in the attempt to take Ticonderoga from the French. One of Samuel Cook's sons was Warner Cook, who married Deborah Robbins, a young school teacher.

She had made her entrance into Ticonderoga in 1805, her worldly possessions contained in a pillowcase slung across the palfrey on which she rode. Her esquire was an old hunter who served the community as school trustee. He walked ahead leading the horse, for the road was full of holes and fallen logs.

In 1812 Warner Cook and his wife Deborah were blessed with a son, William H. Cook, who was the author of most of the following letters.

At the age of twenty-five William H. Cook married Merrette Lamb from Hague on Lake George. On January 26, 1838, their only child was born. They named him Flavius Josephus, a name he was to find too burdensome. When he was a student at Harvard he wisely changed it to plain Joseph Cook. As Joseph Cook he became known the world round—Joseph Cook, the Boston Monday Lecturer.

When he was a little boy his father showed him a barrel of quicklime. "This," said father to son, "is my public orator." He applied water to the lime and pointed out how a low rumbling sound was produced, the contents of the barrel growing in agitation until lime and water became a seething, boiling mass. He bade the boy imitate the lime barrel and become a public speaker to agitate the world. In later years when Joseph Cook addressed thousands every Monday noon in Boston, great crowds on lecture tours throughout the United States, and multitudes in Great Britain, India, China, Japan, and Australia, his father had some reason to believe that the lesson of the lime barrel had not been entirely lost.

The first of the following letters was written in 1851. At that time the village of Ticonderoga was made up of two parts, called the Upper Village and the Lower Village. Neither was more than a trading center for farmers, except on Saturday nights, when the less austere members of the community drank grog and played cards and sometimes held a dance at 'the corners.' In these pastimes their neighbor William H. Cook did not join.

Communication with the outside world was by stage unless it was in the summertime, when steamboats plied the waters of Lake Champlain between Plattsburg on the north and Whitehall on the south. Whitehall was the terminus of the railroad from Albany. In winter the twenty-five-mile journey between Whitehall and Ticonderoga was made by sleighs driven over the frozen surface of the lake.

The mills later operated by the waterfall at Ticonderoga, on the outlet of Lake George, had not yet sprung up. The harvest of lumber, the first crop on the farms of that vicinity, was about at an end. Farmers were turning their attention to livestock. The sheep industry was receiving attention, and Blackhawk horses were in their pristine glory. The best horses and sheep came from "over the Lake," that is, from Orwell and Shoreham and other prosperous Vermont towns which had been settled earlier and therefore were greater centers of wealth and culture. At that time there was no newspaper in Ticonderoga, no schools but district schools, and the schoolhouses served also as churches.

Flavius Josephus was sent away to school when he was thirteen years of age. First he attended Newton Academy at Shoreham, Vermont. After a few weeks at Shoreham he was sent to Whitehall. Besides being a transportation center, Whitehall, it seems, was also a seat of learning. At least there was an academy there for youths and maidens in search of education higher than the district schools supplied. It is on the deck of a Lake Champlain steamboat that we first meet Flavius, a serious child, bound for Whitehall and the protection of Minerva.

Thus began a schooling that was to continue for twenty

years at Phillips Andover, Yale, Harvard, Andover Theological Seminary, Halle, Leipzig, Heidelberg, and one or two other German universities.

The following letters have been edited to a slight degree, chiefly by the insertion of periods and capitals to mark sentence divisions which might not otherwise be clear. The paragraph structure of the original letters has also been disregarded in many instances inasmuch as at least one of the writers seems occasionally to have been guided by the ink capacity of his pen, rather than by his thought, in starting new paragraphs. In general, however, the printed letters are very close to the originals. I am greatly indebted to members of the staff of Cornell University Press for their assistance in the editorial task.

I have not embellished the letters with copious notes, in the belief that little or nothing is needed to make the text clear to the reader; however, a word should be said concerning some of those whose names are often mentioned. Clayton De Lano was a Ticonderoga boy who grew up with Joseph Cook and became a prominent paper manufacturer. He moved to Boston where he was at the head of a large business. A monument to him now stands in Ticonderoga. Hiram Kimpton was another of Joseph's contemporaries in Ticonderoga. They went to Yale together. Kimpton had an unfortunate career as a carpetbagger after the Civil War. Louis Beaudry was a schoolmate of Joseph Cook's before his college days. Beaudry entered the Methodist ministry, served as a chaplain in the Union Army, and was held captive in Libby prison.

Elder Grant was an evangelist in whose family Joseph boarded at Whitehall. Colonel Calkins and Judge Burnet were leading citizens of Ticonderoga. Joel Holcomb conducted a tavern there. Weed and Shattuck are old family names in Ticonderoga. Rebecca Griswold was the wife of an Orwell farmer. Left an orphan, she was brought up in the Cook family. From her daughter, the late Mrs. Ellen Warren of Orwell, I obtained these letters.

One of Joseph Cook's classmates at Yale about whom much is said in the letters was the Rev. William Henry Harrison Murray, whose name has gone down in history as Adirondack Murray.

As for others named in the letters, let the reader remember the custom of the playwright in designating his minor characters. For Elijah or Harmon or Robert, read *A Hired Man;* for Loret, Rowena, or Kitty, *A Young Girl;* for Mr. Wright, *A Parson;* and for Charley or Uncle B. Peterson nothing more than *First Citizen* or *Second Citizen* is required. Little difficulty will be found in assigning to each the part he plays.

"Be satisfied that something answering to them has had a being. Their importance is from the past."

Glens Falls, N.Y. F. G. B.
September, 1945

Letters of a Ticonderoga Farmer

CHAPTER I

As the Twig Is Bent

FLAVIUS JOSEPHUS writes back home after his journey by water from Fort Ticonderoga to the bustling village of Whitehall, where the railroad runs, the lake ends, the canal begins, and souls are to be saved among the boatmen.

VOYAGE TO WHITEHALL

September 14, 1851
Sunday Afternoon

Dear Parents

Begining at the begining I give you an account of my coming to Whitehall. Soon after leaveing the Old Fort I was walking around on the boat and to my surprise I met with Edwin S[h]attuck going to Glens Falls so that he and Mr. Rising were company for me all the way. We had a very pleasent time and ride. About a mile below Whitehall or half a mile below the elbow is the Depot; here Edwin and the greater part of the passengers stopped to take the cars which run from this place up through the vilage. I and Mr. Rising stood upon the front deck of the boat and here for the first time I saw the cars move off. After which the Saltus proceeded also and passing the elbow landed myself Mr. R. and the remaining passengers at Whitehall on the west side of the lake.

The minute I slipped [stepped?] on the dock I heard, "dont

you want that valise carried up" I answered "No Sir" and along with Mr. Rising walked along up canal street. We soon found a man that knew the way to Mr. Grants and bidding Mr. R. good bye I took the mans direction to Mr. G's. I had not gone far before I met Mr. Grant right in the middle of the street. He knew me and said "On hand then be you" "Certainly" said I. He then told me where to go and pased down the street on some business which he said he had to do.

I went along. Found his house, found Aunt to home. I had some supper. Went to bed in Mr. G's study. In the morning I was waked up early by Mr. G's coming into the room after some tracts. He said that he wanted me to go down to the boats with him, which I did, his business being to distribute tracts among the boatmen. He says he shall have me so to distribute tracts to them in two or 3 weeks.

After this, breakfast; then sunday school. . . . I went down to hear him preach to the boatmen. It is astonishing to see what attention they give to all that he says and after it is over they all crowd around him, pressing him to give them a tract.

Monday morning about 7 oclock I and Mr. Grant taking with us my books went over to the academy. He introduced me to Mr. Reynolds the teacher. He looked at my books and said the gram[m]er would do and then wrote on paper the books I should want which were Parkes Philosophy, Thompsons Higher Arithmetic and Thompsons Days Algebra which made three new books which we bought that morning. Cost of whole 180 cts. I am in the first class in Arithmetic 1st class in Philosophy 2nd class in algebra also third clas in Algebra. He thought it best for me to study bookkeeping avery usefull study and he put me also in the gram[m]er class. All of which studies I have kept up with as yet.

So you see that with 5 different studies and an able, efficcient, thorrough going, teacher, I shall be apt to learn something. He wont leave anything behind if it is not fully under stood. He says that in studying philosophy, you should study it, see it, and feel it, and in accordance with this plan he brings out the apparratus and tries all the experiments before the class. We are studying electricity so he brought out the electrical machine. After trying many experiments he let us all receive the electric shock the class first and then the whole school. Such studying makes the school very plaasent.

He has about 60 schollars and more a coming in all the while. Although he has two female assistants he has no time to lose. We have the Bible read and prayers in school every morning and afternoon. I dont think I could have gone to a better place. I have seen Mr. Jones 3 or 4 times talked with him. I have been in good health up to the present time not a sick day nor hour. I like living here better than I expected.

I have not much room now, butt I must ask some questions about home now. Have you all been well? How does the colt come along any better worse or dead? Have you had any rain there? It has been very dry and warm here since I came excepting yesterday and today when it is cold enough by reason of the north wind. We have one rainy day. Are the neighbors all well? Has Mr. Miner come to build the barn yet? How does the union store get along? Give my love to all the family and surrounding freinds. Your obedient son

Flavius J Cook

Whitehall Sep 14th, 1851

The "Francis Saltus," aboard which Flavius voyaged to Whitehall, was built at that place in 1844 by Peter Comstock and a few years later was sold to the Lake Champlain Transportation Company.

A WORD FROM HOME

Ticonderoga Sept. 19th Evening 1851

Dear Sir

We recd your letter last Monday & was glad to hear you were in good health & well pleased with the school. I hope you will make good use of your time. We are all well here. Things go on about as usual here. We have sowed our wheet, cut up the corn and are now fixing for mooving the cornhouse. . . . We have had no rain since you left to speak of. Some have to fodder their cattle here. The colt is dead. I opened the bunch. It prooved to be a breach. He died last saturday. Rebecca came home the next day, after you left. She left 17th for Orwell. They are having a ball at Levy personses this Evening. . . . Your mother will be at Whitehall next week or week after. She will fetch cloth for a pair of pants for you. We should be glad to hear from you as often as once a month and oftner if anything should happen. Give my Respects to Elder Grant and famely.

From your affectionate father

William H. Cook

SPANISH MERINOS

Wensday Morning
[October 12, 1851]

Dear Flavius.

. . . Well, I have got some news to tell you. Your Pa, is getting to be a dreadful man. . . . Don't you think that he went off with uncle Valorous, very early last Saturday morning (intended to have returned the same day) but did not till Monday towards night. Left no one here, but Lorett and I, to take care of the Horses in the barn, do the Milking, fodder the Cows, feed the hogs, and so on. So don't you believe that he is getting to be a bad man? But we got along well enough. We went and got Asher, to come, and helpe us. . . .

Well, I suppose you will want to know, where he went, and what he was about. They went to Vermont to buy sheep. Your Pa, bought two Sheep and paid 30 dollars a piece, which made 60 $ for 2 sheep, and 8 [sheep] that he paid 7 dollars each. Uncle Valorous bought 10 at 7 dollars each. They call them the Paular Merino pure blood. . . .

From your Mother
Merrett Cook

N.B. I must tell you (though I like to have forgotten it) that Pa has bought a one horse Waggon and a Sulky. He says the Sulky is for you to break colts in. Wonderful!

These sheep were among the first lot of Spanish Merinos imported into Ticonderoga. The state of Vermont was then accounted the best wool state in the Union. Two years later the industry was thriving in Ticonderoga to such an extent that the farmers held competitive sheep shearings. The heaviest fleeces were clipped from the flocks of William H. Cook.

He was also a leader in horse raising. In this branch of livestock Ticonderoga excelled long before its farmers went in for fine sheep. Old Mike of the Messenger breed had been imported from Ohio in 1834. Other celebrated trotters had been brought from Long Island. The original Blackhawk was owned at Bridport, Vermont, and his descendants were in high favor with Ticonderoga horsemen. A son of Old Blackhawk became the most celebrated horse in the United States—Ethan Allen, whose dam was William H. Cook's Old White Mare. The letters contain many references to his experiences as a breeder of fast horses.

"WATCH OVER YOUR FALTS"

Ticonderoga January 5th 1852
Evening

Dear Son,

Strive to make as little trouble as possible at your bording place. Watch over your falts. Strive to make thing[s] pleasant around you. Pay good attention to your manners and go ahead. . . . Ceep good company, don't be out Late Evenings, attend meeting & Sundy School. Strive Eventually to be among the top rounds of the Ladder. Be prudent as to expences.

Yours affectionately
Wm. H. Cook

HORSE TROT ON THE ICE

Ticonderoga 18th February 1852

Dear Son

All well here. The fine woold Sheep are doing well. Your colt is doing fine. Drove her a little. Went verry well. . . . Driving the Sorrel. She goes verry well. There has been quite a gathering on Lake Shamplain this week commenced Monday Afternoon. Purse of two Hundred dollars between J. Holdcom & Southern Horse, trotting one mile. Southern Horse beet. Tuesday afternoon purse of twenty five dollars to the fastest trotting Horse. Holdcom win the purse. To day they trot for a purse of fifty dollars. I have been down to the trot both days and think of going today if it dont Storm. . . . Colonel Cooks black Hawks run away on the ice the other day. Not much Hurt done. . . .

Your affectionately
Wm. H. Cook

NO CHANCE TO SEE ANYTHING

March 3 Wensday morning

Good morning Flavius: . . . There has been two gentlemen by the name of Collins from Whitehall staying here since Monday night. They were buying cattle and sheep. Your Pa has sold them fifty wethers, two of the best cows, the steers that he bought of Eli Bowdry last fall, and some yearlings I believe he paid him $270 dollars for. . . .

I should think you would want to go to some of the Great Dances they have there, just to see them. Joel Holdcomb and daughter have been down there to one. It was when the Fire Company and Odd Fellows and all those came out. I should think it would be quite a sight to see them when dress'd in uniform. I would go in sometimes when they have some such great doings, just to see the differance in society, and then you can tell us some thing about it, for you know we don't have a chance to see anything here. . . .

From your fond Mother,
M. Cook

PLANS FOR NEXT YEAR'S SCHOOLING

Ticonderoga March 25th 1852

Dear Son

. . . As your term is nere closing you had better ask Mr. reynolds [principal of Whitehall Academy] what studies you had better pursue if you Should attend School in Some other place. Ask him what place he Should prefer sending you. Ask him if he thinks it best for you to spend a year in canada to learn french. . . .

I should of been very glad to have been at White Hall at the closing up of your school but as the roads are now it will be verry bad going So that I cannot attend. . . . Pleas write in your next how mutch Money you need to close up your consern.

I hardly think the ice will be out of the Lake by the tenth but it may. Pleas write what day you will leave So I can go after you. . . . Ask Mr. Reynolds if he thinks you better continue your Studies this Summer or lay by for a few months. . . . Be careful and not be drawed into any bad company.

Yours affectionately
Wm. H. Cook

N.B. Flavius, . . . if you can get time I want you to go to the Carpet Factory and inquire on what condition they will change rag carpeting for good rags already cut. My old carpet is worn out and I talk some of trying to get a new one. . . .

Well, your Pa and I have been up to Mr. Garfield's. . . . We went up by land, and came back on the ice. . . . We saw a very large PANTHER. It was killed by Jabez Patchin. It measured 7 feet from the tip of his tail to his nose. He was a frightful looking creture, I assure you. . . . Jabez shot at him 7 times before he fell. . . . I suppose he could kill a man as quick as a cat could a mouse. . . .

Your Ma Forever dear.

The following September Flavius entered Keeseville Academy at Keeseville, New York.

HIGH COST OF BOARDING

Keesville Sep 11th 1852,
Saturday.

Dear Parents:

. . . We started from the Fort about 12 and arrived at Port Kent at 4. Fare on the boat $1.12½. Dinner 50 cts. I had a very pleasant ride finding 2 of my old schoolmates at Whitehall on board going to Platsburg. From Port Kent we had a suffocating ride over the plank road 4 miles to Keeseville with 21 passengers in a close omnibus. Fare 25 cts. At the first tavern we stopped at namely the Adirondac[k] Hotel we inquired the price of board and what was our surprise to hear $3 named as the price. *Of course* we did not stay there but proceeded at once to the Ausable House Where we after some bantering got board, a good room, lights, and other conveniences for $2.25 cts. a week, washing *not included!*

"Well, you've done it now, I should think! $2.25 cents for board, washing not included! Why didn't you go to some private family and get board for $1.75?" I think I hear you say as you read this. Well, I will tell you. Simply because I could not. Board and everything, books and all are *very* high here. . . . Although I have nothing to find fault with except the price I would not have come for the benefit I shall recieve I am afraid will not be equal to the expense. . . .

How does horse racing and the Sorrell prosper? How does the harvest, and the crops turn out? Have you all read uncle Toms cabin yet, and how do you like it? Has pa read it, and does he call it "a little boys story" yet?

From Your Affectionate Son
Flavius J. Cook

EXTEMPORANEOUS SPEAKING

Keesville, September 18th, 1852
Saturday Evening

Dear Parents:
. . . Since we have been here I have made Lewis [Beaudry] join me in extemporaneous speaking; and we have been sevral times to said grove, to deliver our speeches, which, some of them are not "to be laughed at," no more than were Henry Clays orations to his fathers cow and flock of sheep. . . . It is easy for me to take any subject and speak 15 or 20 minutes on it. . . . I made one speech on slavery, 20 minutes long which I wish pa could have heard. . . .

Flavius J. Cook

THE BOY ORATOR IN TRAINING

Keeseville Nov 6th A.D. 1852
Saturday Noon

Dear Parents:
. . . Yesterday noon I received my things all safe and sound, *sugar, apples, cloves* and all.

. . . I must tell you a little about our exercises last evening. I had the most to do of any one; in the first place I spoke a piece on the First Prediction of an Eclipse by Proff. O. M. Mitchel. Next I read the gentleman's compositions, and lastly I acted the part of an Indian chief in a dialogue on the Indian's Wrongs. . . . When we left the stage and started (as the dialogue reads) for "our western home" the School choir struck up

"Wild roved the Indian girl
Bright Alpheretta,
Bold was her warrior chief," &c.

This produced a great effect. . . .

Your Affectionate Son, Flavius J. Cook

After sending his son to Keeseville to school, Farmer Cook decided that the boy's horizon should be still further extended by studying in Canada, to which place the next letter went.

A WINTER ECLOGUE

[Undated]

I think of you Some of these cold nights. I hope that you are as comfortable as you would be here in your little room with Ma to find cloths for to ceep you warm. But if you Should always be fishing in a milpond you would never know how large the Ocean was.

I am writing here on the table before your books and case. They are all aranged verry nice. Ma is reading Mr. Judsons life. I read in your books Some but I am quite tired and Sleepy in the evening after being out in the wind all day. I am lazy, heedless, and negligent as regards reeding. I Spend my time principly in thinking.

As I now rais my eves from my paper it [!] fell on your papers that lay in your case. The first thing I Saw was an adress to the Caball. I hope that I may live to See addresses from you that will compare with a Websters or a Franklins. . . .

Your Afate Father
Wm. H. Cook

GOOD REPORTS OF FLAVIUS

[Undated]

Dear son
. . . I returned from Elizabethtown last evening about dusk. I Saw Some of your K[eese]ville friends at court. They Spoke verry highly of you and your doings while there. . . . Spoke in particular about your Speeking at the S[unday] School celebration. Said it was verry good. Spoke of your Speeking at other times and other places. Said you was a great Speeker for one of your age. Says K[eese]v[ill]e never had your equal before nor Since. . . . This is Souch news as I Shall hope to hear from all the places that you leave. You must bear in mind that it is with you as it is with all others that have gon before you, wherever you can See or hear of ones taking the lead, advances new thoughts, new ideas, and new things, you will find the worlds people to follow him like the drift wood of a mighty river in the time of freshit. Therefore it Stands you in hand to try and take the lead and Strive to be the one that is looked up to for new things, new thoughts, new ideas. . . .

I hope you may be able to learn french quicker and better than any other yankey Rep[resenta]tive that ever entred Cda. . . . I think you had better Stop at Andover a year or more So as to get a K[eese]v[il]l[e] Standing. . . . What has been done by a Webster can be done by Some one else, if they have confidence and Strive to do it. Dont fear the face of clay for in ninety nine cases in a hundred you will find more on the boddy than in the head. . . .

Pleas leave your postage unpaid as the Pmaster Sais it is the Same here as there. There is no advantage in pre paying.

Your Afate Father

Wm. H. Cook

Elizabethtown, where William H. Cook heard so much in praise of his son, is the county seat of Essex County, New York. Farmer Cook was often called there—as a juror, as a commissioner of excise of Essex County, and once, as we shall see later, as a litigant in a celebrated case.

"SOAR ABOVE WASHINGTON"

[March 7, 1855]

Dear Son:

. . . I think that by the time that this letter reaches you, you will be able to talk the french Language quite well. I think as you become familiar with the language you can learn it faster. I hope you will strive to learn a great many things while you are there. Make the island of Montreall as familiar as posible. Learn all about the people, their manners, customs, ocupations. It would be well to ceep a little journal of the advancement of Spring, the appearance of the first Spring Birds, the first green grass, the putting out of Some of the forrest trees, the time of their commencing gardning and of plowing and Sowing. Anough of this at present.

I am unable to express in words my feelings for your future welfare an greatness and goodness. I hop and trust that you may Soar above a Washington, a Webster or Ward Beacher. But Gods will be done. . . .

I see by your last letter you were about commencing your Speeking or oritory in french. I hope you may be as Successfull in your Speeking as you were at K[eeseville]. I hope you will pay particular attention to your oratory. Strive to get as near perfection as posible. You must Strive to lay up a granery full of knowledge for future use. . . .

Wall, I suppose by the time this letter reaches you, you will be back into your old habbit as regards exercize, Something Similar to the woodchuck or the bear thro the winter. But perhaps it may be otherwise. It Should be after all that has been Said. It [is] our duty to attend to our health. We were maid for action and if we neglect it we become as a thistle that grows in the cellar with drooping head and Stooped form. You know better than I perhaps what is necesary for you to do to preserve your health but I fear you lack energy to carry it out. But anough of this for the presant. It is an old tune to you.

I would Say to you as regards Speeking and debating and conversation, go amongst the big Swells far out from Shore where the white caps break and foam, where the Swells roal Mountain high, and learn to paddle your own canoe. Dont fear the face of clay. Learn to Stand up with the best of them, always Showing respect for your Superiors. You must have an eye to your own interest. If there is a Spot in canada more favourable for you to learn french than the one you are in, you had better find it. I think you will have to Stop a year at Andover before entering Coledge, but this we can talk over at Some future time. . . .

Your Afate Father
W. H. Cook

Sunday A.M. March 7, 1855

Ever Dear Child,

. . .Your Pa and I have been to meeting this forenoon. We have been to meeting considerable for us this winter, and we find the more we go the easier it is for us. And next sabbath he preaches a surmon to the young men. And your Pa says to Charley since we came home that they must rig out the double sley and all go. So you see that he is more engaged, than usual, about going to meeting. So take courage about us and not lament too much. . . .

Now I will tell the general news. There has been a fire in

the village the last week. Mr. Fraziers work-shop, with all its contents, valued at about 500 dollars, he had a set for our bed room almost done in it and that was burned. . . . Old Mr. Sheldon is dead. Mr. [W]right preached his surmon from the same text that Mr. Havens did your Grandpas. Mr. Havens preached to the village last sabbath, and there was a Catholick meeting also, so there was a great turn out. We went to the Baptist Church. . . .

I feel to love Madam Vanier for her kindness to you. Give my thanks to her, although she is a stranger. Your Pa thinks that you will be apt to form a habit of drinking tea, that will not be so easy dispensed, but you must be your own judge. If you think your health is better by using it, I would while there. Do watch your health. Exercise the body and not the brain too much. You seem to have no fears on this point but I have more fears about that than anything else. But we must trust to the seeing eye. If he has a field marked out for you, he will prepare you for it.

From your Mother,
M. Cook

FARMER COOK FEARS FOR HIS SON

Ticonderoga April 29th /55 Sunday Eve

Dear Son

We have been to a funeral to day. Mary Rogers, after a Short ilness, has paid the debt that we all have got to pay. I hope and trust She is better off than the Liveing. The Services were performed by a methodist man . . . , from Job 14th, 10th verce. This is the fourth Sermon I have heard preached from the Same text within a Short time. All constrew the meening verry differant. All Seem to be quite positive that their way is right. Its a matter of indifferance to me. I put my trust in a higher power than man, hoping and trusting in the great ruler of all things. . . .

I was in hopes you might be fortunate anough to of got into a famely where there was a good and business man, one that you could of patterned after, but as it is otherwise ordered I hope for the best. I See by your letters that you are all Religion, all Mishionary, all Meeting, all good. I think you are in a place where you can enjoy all of these things. I am glad it is So. We have all of us got a mishion to perform. The great ruler of all things will guid as it Seemith to him good. I would advise you to pay Some attention to business. I am afraid you are agoing to make a Sort of a Softly femenine man. I Should be Sorry after Spending So mutch of my heard earnings to educate you to find at the expiration of your Studies that you was not capable of getting a liveing. I Sometimes fear it will be So, if you pursue the course which now Seems to be your favourate theme. You have got many a dificulty to encounter, a hard battle to fight, but gods will be done.

You Say that things go well and Smooth with you. You think you are fitted to take things easy and pleasantly by their Smooth handle. Was there ever any other handle presented to you but the Smooth handle? You have gon thro life as it were in a quishioned rocking chair bolstered up as it were with downy pillows and drawn by a good team. All you have to do is ask and it is given. You Should think of these things for the time is comeing when you will not be thus provided for. . . .

I have lost a good many Sheep. The prospect is not verry favourable for the farmer. Wool is verry low, provision is verry high and hard to be got. You must be prudent of your money. You have already Spent more than we calculated it would take to carry you thro. Cant you get into Some business So as to part pay your way? It is hard for me to have to borrow the money for to ceep you along. . . .

Your Afate father

Enclosed I send you $15 which you must use prudently and not give it away.

Wm. H. Cook

HARD TIMES FOR THE FARMER

Ticonderoga May 27th 1855

Dear Son,

. . . I See by your last letter that you think the chance is verry favourable for the farmer. I will tell you how we are Situated in this vicinity. We have had but two Slight Shours Since the 15th April. It is verry dry. The fields that have been Sowed for months Show no Signs of green at the preasant time. We have cold north wind for 15 days in Succession. While I am writing the wind is blowing from the north a perfect gale. The fires are out on all Sides of us and it is quite Smoky thro the valley. The grass is hardly good feed for cattle in the best of our meadows. I never Saw things so backward. . . .

The Stalk is doing tolerably well. I have a fine lot of lambs. The colts are doing well. The horses have got verry poor from hard work this Spring. We want rain to Start the grass. If we dont get it Soon we Shall get nothing this year. Apple-trees are in blow. . . .

I see by your letter that you talk of taking a tramp this Spring. I cannot give advice as regards the idea. . . . If it can be done with Safety to your health and person, I think it a fine plan. . . . I have a few words of advice, Supposing you Should go, and they might not be out of place if you Should not. In the first place, you mus consider the cost. I am told they charge one dollar p^{r} meal on the Steam boats. If So, it would be rather expensive board. . . . I think it would not be prudent to be to noisy about Romanism or any other particular Sect or even be verry noisy about anything as you are so young and from the States. . . .

Inclosed is fifteen dollars which aded to the amount you have used up Since you left this place amounts to $129. I hope you will be prudent as the money I Send is borrowed.

If the weather Should become Sufficiently warm I Shall Soon wash and Shear my Sheep. If not, I shall let them wear their fleeces at preasant. I will rest now for a while and take a walk over the farm. Things look gloomy & desolate. . . .

Lewis B[eaudry] has been in this place for about three weeks. He has preached two Sermons at the School house in tuppertown. He has had a full house both times. . . . I heard the first Sermon. . . . He has a great many of the methodist cants & frases and ways. I think he wil be liked verry well. . . .

Your Afate Father

Wm. H. Cook

Tuppertown, more properly Toughertown, was a section of Ticonderoga where the stubborn, stony soil caused the early settlers to say that they had to "tough it."

ANOTHER PARENTAL ADMONITION

Ticonderoga June 18, 1855

Dear Son,

. . . The rain has fell in abundance, since I wrote you last. The flatts have been flowed and some damage done to the crops. Our corn looks very bad. Have got to plow it over and sow it to Buckwheat. But this I will not find fault about. I will do my duty. The prospects are very slim for the farmer, . . . but there is no use of grumbling. It is all for the best I suppose or it would not be otherwise.

About your anticipated journey to Quebec, it will be for your advantage to go, I *suppose* wether it will pay what it cost or not, I cannot say. It is dear traviling these days, and you are not calculated to be quite as prudent as some. There are some young men that would take that trip for half what it will cost you. You do not [know] the worth of money, as you never had to make any exersion for it. . . .

Prehaps you had better go it if it can be done with safety to health and boddy. I hope you will try and learn a great deal in your tramp. Dont be in a hurry, dont be afraid to talk and ask questions. . . . Dont leave Canada till you are perfectly satisfied that you can talk correctly.

The money that I send you is hired. When it will be paid, I cannot say. I have sold nothing of any amount since you left. I have my help to pay, family to suport, money to send you, all done on a borrowed capital. . . . Wool is very low at presant, and everything else, that I have to sell. Haying coming on, and no money only what I borrow. The future looks dark to me. . . .

In this letter I send you fifty five dollars, five dollars more than you called for—it is in large bills as you see. I hope for my sake you will be very particular in making change. I hardly think that you care enough about money, to be particular in making change, taking what is given you back without any regard to counting. Amediately after receiving this money, I want you to pay your debts, before you sleep. It is astonishing to me, when you are receiving money in every letter, that you don't keep paid up. I hope you will for the future. Don't buy any more *books*. They will [be] worthless to you hereafter. I want you to bear in mind, that if you live, you will have to paddle your own canoe, some future day. I now bid you good buy, hoping that you may be prospered as well in the future, as you have been in the past. . . . Your teacher wrote favourably of you. Give him by thanks for his trouble.

Your affectionate father,

Wm. H. Cook

CHAPTER II

Andover in the Fifties

As THE READER has already learned, while Flavius was in Canada his father had decided that he should "stop a year at Andover."

ENTRANCE TO ANDOVER

Andover, Mass. Mansion House parlor,
8 o'clock, Wednesday Eve., Nov. 14, 1855

My Ever Dear Parents:

When Ma [mother pencilled in above] said good-bye I saw she was afraid for me & I was almost afraid for myself. At Rutland (fare 95 cts) I met Mr. Rowley going to Boston with cattle on the cars, & Mr. Baldwin a young man I knew well in Keeseville. Neither of them however were going my route. We passed Bellows Falls about sunset, Keene as it was growing dark, & finally stopped for the night (fare, two hundred miles, $4.10) at Fitchburg in Mass. My baggage had been marked for Lowell but I let it go on & didn't fret and slept soundly. Had a good supper & breakfast (for $1.25) & was off again as the sun rose. At Groton Junction I found my baggage & found myself with it about noon at Lawrence, near Andover . . . , & at this place my name on my box introduced me to Mr. Jacob Allen . . . who came up, inquired after my family & all the older people of Ti. Said he was born there, invited me to his office, showed me over his mills, talked politics, & drove me back to the depot, from which place I was set down at Andover, all safe & sound about one o'clock P.M.

I inquired my way straight to the house of Mr. Taylor, the principal of Phillips Academy. I found him at home, & must tell you that he looks just as Judge Garfield did, just as fat in the face & sides, a little broader & taller & probably more wise & talented. "I see Mr. Taylor, do I not?" said I. "I am Mr. Taylor," replied he in a low & easy tone and with a kind manner.

"I am a young Man from New York State who wants to study Greek & Latin at your Academy, & who wants to find a place to board where I may pay for it with the work of my own hands, & I am come to ask you what I may do."

"There are many places of that kind already filled," said he, "but you will do well to call at *Hon. John Aikens*," & he directed me to the house. . . .

I ate dinner out of my bag at the depot, read the papers, walked around the village, which is very large & open, full of elm trees, white clean houses, beautiful grounds, & young ladies & young men walking the streets, appearances of quiet thrift, good people, & intelligence, all on sandy soil with few mountains or hills in sight—and a little after six I was sitting in Mr. Aikens' parlor with him.

"Ah!" said he after I had told my wants & the name of my own town, "Ticonderoga! I know a man there of some legal note who used to be one of my Scholars."

"Judge Burnett," inquired I.

"Yes, I was his tutor in college."

"I have a letter of recommendation from him." . . .

He would have me take supper with him & I was thus introduced to his wife . . . & I left with the direction to call tomorrow at 4 P.M. . . .

I shall go to the academy to-morrow morning & study & work hard every day. This voyage has profited me much. I have thirty pages in my note book written on the way, a careful description of all I saw that I thought worth observing.

I find on adding all up that since leaving home I have expended *eleven* dollars, ten of which has been for railway tickets & lodging. I am looking out for Mrs. Stowe & other great minds in this place. I see Daniel Webster's portrait hung up everywhere. A description of the Academy & its regulations I must leave for my next letter. . . .

Flavius J. Cook

ON THE WAY TO FAME

Ticonderoga Saturday Evening
Nov 24th /55

Ever Dear Son

. . . I am of the opinion that you are on the broad highway to fame now. I hope between both our efforts you may be able to go thro Phillip Academa & Yale with honour. If we take it cool, we Shal accomplish all. We must not expect it wil be done without a great many chequered scenes and many a gloomy hour and many a hard days toil, many a tear shed . . . —perhaps dishonour, abuse, Slander, and all Sorts of persicution, but it is our duty to ceep on, trusting in the great ruler of the universe. . . .

I think you wil find a differant School at Andover than you have ever attended before. You wil not find that flattery that is used in Schools where they are Seeking for more Schollars. You wil not get the praise there that you have been in the habit of receiving. It is best that you Should not. Work on do your best, and all wil be wel.

I have a word to Say as regards your attending Lectures. You must not do it to the detriment of your Studies. Your knowledge Should come up like a boiling Spring to overflowing with new ideas, new thoughts, new things, not poured into the mind as into a hopper or bin, to slide of when it is ful and cloyed, a mear recitation of what Some one [else] has Said. . . .

You must ceep me informed from time to time how mutch money you want as it takes about a fortnight to exchange Letters. We Shal cil hogs next week probaly. I heard from Lewis [Beaudry] to day. He is teaching . . . , preaches Sundies and Some evenings . . . gets 23.75^{cts} pr month and boarded. I almost wish you were doing as wel. . . . I have heared mutch Said in favour of your temperance Lecture, but praise is worthless. . . .

Your Afectionate Father
W. H. Cook

Enclosed I Send you ten dollars.

This temperance lecture was a discourse entitled "Intoxicating Beverages, their Composition, Adulterations, and Physical Effects." On his twentieth birthday Flavius delivered this broadside before the citizens of Putnam, Washington County, New York. The same year he spoke on the subject from the platform in Ticonderoga, Port Henry, and Crown Point, all sister towns in Essex county. Judging from certain observations in the letters of William H. Cook the moral effect of this lecture in Ticonderoga was of brief duration.

FLAVIUS AS A BOOK AGENT

Andover, Dec. 13, 1855
7:00 P.M. Thursday

Ever Dear Parents:

. . . I have actually earned $14.05, since the date of my last letter, at the book-selling business. . . . I went to Boston where I was met by the young man who employed me [through vacation], Mr. Hammond, who gave me $3.00, which gained my confidence, & then sent me on to Dorchester, a large town just out of Boston, to visit the schools. My business was to offer *"Nason's* School-Room Songs," & "Noyes' United States Writing Book" for sale or in exchange for *"Fitz* Sc[h]ool-Room Songs" or "Payson & Dunton's Writing Books." The song-book we sold for $0.25 or exchanged for Fitz, no matter how much worn, for $0.08 & the writing books we sold for $0.87½ a dozen. . . . I was to leave specimen copies, take orders, & keep moving.

Well, that day I saw Jonathan Battles, Robert Vose Jr., B. F. Brown, & visited Isaac Swan's house, all of which gentlemen are teachers of large schools. They received me kindly; I was pleased with the class of men I came in contact with; they showed me their schoolhouses; gave me much information as to the habits & manners of their educational system; but as their schools were not in session . . . they could not tell whether their scholars would want the books or not. I left a copy as a specimen with each. . . . I was joined next morning, Dec. 1, by Mr. Hammond, & we rode to Quincy, a large town eight miles south of Boston to see the schools there. . . .

The next day, being Sabbath, we walked to Boston on the railroad, through a dense fog over salt meadows . . . to hear the celebrated Theodore Parker preach. . . . I stayed in Quincy until Thursday & visited seven schools, whose masters names I wish to preserve, viz: Stephen Morse, Mr. Paine, C. Murdock, Seth Deering, Mr. Wellington, Miss Frye, Miss Vezy. . . .

John Adams, 2nd President of the U. S. & John Quincy Adams, 6th Pres. of the U. S. were born & buried in this town. I visited, drew a picture, & got some antiquities of both houses, now 154 years old, where they were born. I have a drawing also of their busts in the old Stone Temple Church. These drawings I sketched before the schools had begun in the morning & finished up nicely in the evening. . . .

But my books did not pass readily. . . . I was obliged to return to Boston. . . . I visited the Phrenological Rooms of Fowler & Wells. Here 9 pages were written for me in description of my head & advice, which with the other information

I received, I think worth more than $3.00 which I paid for it. . . .

Your Affectionate Son,
Flavius J. Cook

The drawing of the Adams houses referred to above is in the possession of the editor.

"ONE THING THAT YOU WAS MADE FOR"

Ticonderoga Dec. 23, 1855
Sunday Evening

Dear Son

. . . Strive to help yourself Some as you have now broken the crust in the pedling voige. Don't it Seem pleasant to Spend money of your own earning? . . . Would to God that I was capable of giveing you the advice that my heart wishes you to receive. . . . I would ceep myself wel versed in all political matters the foreign news and the general topicks of the day, So as to converse intelagently with any one. Try to improve yourself and others in all things. Dont fall back in your Oratory. Accustom yourself to Speaking at all proper times. It is one thing that you was made for. . . .

I would be temperate in all things. Don't get into any habit of drinking any hot drink. Let cold watter or good milk be your favourate drink. Dont follow after any of the isms of the day for they are productive of no good. Dont think to mutch of Phrenology. All these things are in the dark. I am told that Fowler & Wells are of the Same class of T. Parker, and that many in Boston concider them greater than god. . . .

Inclosed I Send you $10. . . .

Your Afate father
Wm. H. Cook

Monday Morning 24

Dear Son,

. . . I hope you will take care of yourself and not be sick. I have been sorry that I did not send you some herbs, to use when you have a hard cold, but I did not think. What do you do? You must not suffer a cold to settle on your lungs. You had better get you a bottle of Camphor and some Loaf Sugar for I think that is very good. I have tried it some myself, but if necessary go to a Doctor, or get some thing more powerful. . . .

Remember your Mother
As she ever will you,
Through life untill death
She will ever be true.

M. Cook

A PHRENOLOGICAL DISCUSSION

Enclosed is $20.

Ticonderoga Jan. 7th 1856
Monday Evening 10⁰⁰

Dear Son

. . . As we did not receive your Phrenological Chart until after we had Sent our last letter, I wil Say a few words about it here. I am not a full believer in Phrenological works. It is probaly owing to my ignorance.

In the first place Messrs. Fowler & Wells have become So accustomed to there business that they can get verry near the mark. In the first place they Say you have an unusal Large brain. Anyone must know there is Someting in that large head of yours either Sap or brain.

Your long-live ancisters they told you about must be on your Mothers Side as there has been no long lives amongst the Cooks.

They say you wil grow Smarter as you grow older. Thats astonishing. I hope it may be so.

They Say you will have influence after forty. If you dont I think you may give up that you never will. . . .

They say you wil grow better as you grow older. Thats a headacher, although there are many people that that cannot be Said of.

There are many things that I like mutch in the chart, Soutch as your nature holesail and generous. . . . They say your bump for accumilation, if it means money, is not comeing til after forty. I am Sorry as I Shal get no help to furnish means to get you thro College. . . .

Upon the whole its a nice thing. Strive to improve by it, and grow wiser and better, every day. Use it as a private Lookinglass for your own benefit not for others to make fun of, and here I wil close this foolish Chapter as you wil Say. . . .

Your Affectionate Father
William H. Cook

The years proved Fowler and Wells right in more than one respect, namely, that Flavius would have influence after forty, and that not until then would his bump of accumulation appear. His father supported him until he was thirty-five.

A DONATION PARTY

February 2, 1856, Ticonderoga
Saturday Evening

Ever dear child,

. . . We have none to much work at presant, ride when we choose, go to meeting, go to parties, go to visit all friends, and have company. And then we have a plenty of good books, and papers to fill up the time. . . . Last Thursday evening, we attended a donation party to Mr. Rights. . . . There was a great many there, and I think there was considerable given him. I spose you would like to know what we carried. Well, we carried 10 lbs. butter, 1 cheese, 1 box of honey, 3 dozen Candles, and 2 bushels of buckwheat flower. Mr. Right is trying to do a good deal this winter. He is trying hard to do something in this neighbourhood. He has preached to the school house every Tewsday eve. for 3 weeks, and is to preach again next Tewsday eve.

He brings Old Mr. Edwards with him, and he speaks very fervently. One eve. Mr. R. discourse was to the young people, after which Mr. E. arose and spoke very lengthy. Brought you up as an example. Spoke your name plain. Aluded to your being sneer at here, and how little you cared about it. And then he made a very feeling appropriate prayer, in which he prayed for you, for your parents, (we were both there) and all. And may his prayers be answered. For if we are not in the right, it is time we were.

Sunday 4 P.M.

This morning at 10, we started the boys with Old Bay, and cutter to the village to meeting, and your Pa and I, with the colt, and little cutter went up to New Hague, to hear Lewis preach. There he stood in one corner of that little house with his little congregation seated about him, (only 17) trying to instruct them in the right way. . . .

Your Mother
Merrett Cook

WHEN REPUBLICANS WERE FEW IN ESSEX COUNTY

Ticonderoga, Sabbath, March 2d 1856

Ever Dear Son,

We feel to rejoice with you, in the progress of Your Studies. We are mutch pleased . . . that you have been permited to enter the advance class. . . . I think your advantages to learn to Speak are very great at Andover. . . . Dont Speak to mutch nor to often but when you do Speak, do your best. Let them know you are there.

There is one thing I fear you lack in Andover. That is Refined Society. In this you need mutch cultivation. I hope you wil bear this in mind. Strive to So inform yourself that you can be perfectly at home in the most refined Society. . . .

I must finish this page then away to my chores thro the deep Snow with old bay and my fork. The stalk are all doing well. I have lost but one lamb this winter. The Jonny mare is with fold, old bay is not. The colts all look wel. The potatoes rot bad but there wil be a nough for our use. They are worth but twenty five cents a bushel. Apples about gon. We are getting up wood. I am going to getting Some logs boared for to fetch the water to the barn over the brook, likewise to the barn where Charley lives. . . .

The prohibitionist is not taken here by any one now. Affairs Stand about as they did when you left as regards the Main Law, I think rather worse if anything. The Sons [of Temperance] are a going to do Something verry Soon. That has been the talk ever since you left.

Our town meeting comes of next tuesday. The Republicans are So few I hardly think thay wil come out as a party. I understand J. Burnet is to be the nomination for Supervisor on the no nothing ticket. I did not atend the caucus last evening. . . .

Mr. Gould has had a donation party held at Mr. Tefts tavern. There was a general attendance. I was not there. It seems that Mr. Wright has preached a Sermon on gambling and drinking and other Sins that are commited about the corners, which Seams to have offended them. They Say Mr. wright is medling with that that is none of his buseness, and they flock to Mr. Goulds donation. Would to god I could Say to his meeting likewise. . . .

Your Affectionate Father

Wm. H. Cook

RUMMIES REJOICE

Sunday 6 April 1856

. . . I am verry Sorry to Se Souch a differance in your footting, your expences and the money you have Rec[d]. I Should want to know what became of my money. It Seems there is twenty dollars differance, if you ad the amount that you earned the first vacation. How do you know that it has not been Stolen from you? There is a mistake of a doller in your footing in your last letter. . . . I wish you would try to learn to add a Small colum of figures correctly. . . . I wish you to consider every dollar that I Send you worth as mutch as though you had to earn it chopping cord wood or digging ditch by the rod. Something Simelar to this is the way that I have to earn my money.

I have no fault to find as to your expences except your books. This is always your fault. Your books wil not be worth half price in a verry Short time. I hope you wil dispence with buying books. . . .

Keep out of debt if possible. Do your best at all times. You Say you tried to get your room mate Moses up in the morning. I Should like to See the chap that could lie a bed later than you can when at home. . . .

The colts are all doing wel. The one in the new Stable is a verry fine colt. . . . I was in hopes to of Sold Some colts this winter but I have had no chance. Horses are poor property to rais money out of. . . .

Wal I suppose you have noticed the decision of the court of Appeals that the main law was unconstitutional. The rummies are having a fine time of it here, exulting over the bowl in the downfall of the prohibition law. Let them go it. . . .

Strive to make yourself as wel versed in the affairs of the united States from the discovery of Columbus to the presant time, as I am with my own farm. And at the Same time learn the bible by heart. . . .

Hadant you better have the tribune Sent you? You Should

have a good political paper to lay your hands on at all leisure times, as the affairs of our country are Something Similar to the Slacking of a barrel of Lime, and where they wil cool down or when is quite uncertain. . . .

Inclosed is twenty dollars, the Same old Sum. . . .

Your Affectionate father
Wm. H. Cook

A BUSINESS CAREER BEGINS AND ENDS

Andover, 4 P.M., Saturday, Apr. 19, 1856

My Dear Parents:

I slept every night but two in my own room at Andover during the whole vacation. Soon after writing my last letter, Morril [my room-mate] who was in Boston, got me engaged for a book publisher as an agent to sell "The Life of Nicholas I, Emperor of Russia," & I was sent to this town in selling it, by the orders of its publisher. I offered it to twenty-six men, only one of whom subscribed, so I concluded that the book would not sell & sent him my specimen copy, & wrote to Morril to buy me no more books to sell in N[ew] England.

I had very good success in selling [writing] paper, while I worked at it, but the muddy roads, rainy weather, frost-heaved railways, my laziness & desire to do other things kept me in my room more than half the time. . . . In fact, I did not have the first beginning of taste for the business of a pedlar. I thought in thus walking about I might grow stout, learn human nature, & see many beautiful & useful things. Alas! no; my heavy valise stretched my arms until I could hardly lift them to feed myself. Every beautiful or useful scene as I passed it & every chance for learning man I was blind to; dollars were before my eyes like scales. I was thinking of money not knowledge; & instead of learning others heads & hearts I found myself making my own worse by praising my goods above measure & sundry other traders' lies. Therefore I concluded that I was not made to peddle. . . .

So I have only sold about $2.50 worth of paper. . . . I have lived on bread, milk & molasses, milk, molasses & bread, bread, molasses & milk regularly every day I have been here.

Your very affec. son,
Flavius J. Cook

FATHER HAS SOME MISGIVINGS

$20 inclosed — Ticonderoga Friday May 2d 1856

Dear Son:

We become verry anxious to hear from you before your letter arrived. We were fearful that Something might of happened to you in your pedling voige. . . . You say that your valiece made your arms lame. Verry likely. I have been lame and Soar many a time So that I could hardly turn myself in bed in working to get money to Support you and as the prospects are now I have got to do it a great while longer. . . .

You Say the thoughts of dollars got into your head. This you thought rong. I think it right to think of dollars in an honest and business way Sufficient for our own Support. If I mistake not the bible approves of this. . . . Think of the valece and lame arm whenever you use money. The differance between you and me are this. I think to mutch of money and you to little. . . . At times I am verry Sorry that I undertook to educate you. I dont think it wil ever pay. . . .

We are haveing a beautiful Spring. It Seems to make a mans heart grow big with gratitude and praise to the great ruler of all things when he looks upon the green fields and expanding buds and frequent Showers. . . .

Your Affectionate Father,
William H. Cook

THE GRASS SUITS PA

Sunday 4, 5 oclock, 1856.

Dear child,

. . . I think your Pa takes things rather more quiet this spring than he generally does this time of year. He came in the other day and said, There is one thing I am satisfied with, for the presant. The grass looks well enough. You know that this was considerable for him to say. He troubles considerable about you, gets most discouraged, and says that he does not know how he is a going to get along and keep you in money if you are always a going to use so much and not earn any yourself. Flavius I sometimes think that your Pa would be a differant man, and not be so over anxious for making money if it was not for his ambition to carry you through your education. I sometimes think, that it is all that he cares anything about, even his own health, or mine. . . .

This appears to be his greatest trouble concerning you, that you will not be capable of taking care of yourself, when the time comes that he cannot help you. . . . He does not think but what you live as prudent as you can now, except buying so many books. This you must try to dispense with. And not go without comfortable clothes, and decent healthy victuals, for the sake of buying books. . . . It grieves us very much to think how you live, and how hard you must fare. After all you have so much money. Bread and milk, for breakfast, dinner and supper, or bread and molasses, how can I bear the thought that a child of mine should [live] in that way. Neither is it healthy. Especialy molasses. I advise you to dispense with that, for it is injurious to your teeth, and weakening to the whole system. I will send you an extra dollar to buy something extra to eat in your room, such as a peice of cheese, or butter, to eat with bread, instead of molasses. . . .

Your ever loving Mother.

"WATCH KANSAS"

Ticonderoga May 25th 1856

Dear son

. . . You say you are paying Some attention to your debating Society. Strive to improve in speaking. Be particular as regards your jesture and manner of delivery. These things are teling uppon an audience. . . .

We have brought a fine stream of water to the Covil place. The logs are laid from the Spring to where the old hog pen Stood. The water comes fine. This week we Shal fetch the water to the Sheep barn over the brook. That wil finish fetching water for the preasant. . . .

Ceep yourself posted on all good things both forreign and at home. Watch the affairs in Kansas the doings at Washington. Learn the iniquity in man how they profes one thing and act another. Ceep minuets of all important things. Be prepared to address a community at any time when called on, on any Subject. Do your best in all things. . . .

Your Affectionate father

William H. Cook

$20 in this.

SOME GENTLE SARCASM

Ticonderoga July 20th 1856

Ever Dear Son

We feel very thankful that the great ruler of all things has Seen fit to preserve your health and good Standing since you have been at A[ndover]. I hope you have made great improvement Since you have been there. Of this I have no reason to doubt. I like to hear that you take Some interest in politics. They Should not be carried So far as to injure your Studies. . . .

As regards comeing home I hardly know what to tel you. My

first thought would Say come home the cheepest rout as I have no money to Spare for you to travel at preasant. If the $20 inclosed is Sufficient to carry you either of the routs you mention I am willing you Should go. . . .

We are now farely begun haying. Help is Scarce but you wil Soon be at home to Swing the Sithe and handle the fork. How the hay wil fall and fly by soutch a Strong arm as yours. I shal have nothing to do but look on and carry drink. Wont this be fine times? . . .

Your Father
Wm. H. Cook

HORACE GREELEY AT THE COUNTY FAIR

Ticonderoga, Sept 21st, 1856.

Dear Son,

. . . Charley & myself have been up to the county fare the past week. We had a good time of it. We took up the colt that belongs to him & me. He took the first premium among the two year olds. We took up Mas rug and that took the first premium.

We also heard Some Speeking from all Sides on politics—Messrs. Watson, Hand, Ireland, Simons, H. B. Nothrop & Greely.

Watson & Hand are Bucanon men. I don't fancy them mutch. It don't seem to me that they are honist in what they Say. Mr. grely delivered the agricultureal adress on the ground. He is no great orator. He Speeks verry candid and to the point and no man can doubt what he says. He spoke one hour. Then Mr. Nothrop, the same man that went south for S. Nothrop, addressed the audience one hour from the same stand in favor of the American party. He is a strong man and a thundering speeker but he had to ceep truth on his side as he stood under the scrutiny of a traveling dictionary, as he called Mr. Greely.

He made Some fun of greely and his paper calling him a

fusionist a turncoat and many other things. greely sat just behind him looking over many old papers paying verry little attention to what was said except now and then a smile whenever he straid from truth. Mr. G. would say it was not So then he would correct himself. He spoke over his hour and had to stop before he was thro as the Sun was nerely Setting.

After a few minutes music from the Keesvile band Mr. Greely arose. We gave him three deefning cheers and he commenced. He is a great political Speeker, prooving things perfectly plain by producing documants and reading. He read the two platforms compared and explained them. He was called to read the Bucanon platform but choose not to do So as there was no Speeches to be made for that party. His maner of delivery is easy and affecting, Similar of a father to his disobedient children. No man can doubt the cinceraty of what he says if he hears him Speak. He has few jestures. I presume none that he is aware of.

He is a man abou[t] my hight light complection light hair light whiskers far under the chin. His dress was black brodcloth thick boots, black Leghorn hat black ribon or corde attacht to a gold watch. His walk is rather Slow with a Slight Swager. . . .

It is the opinion of the knowing ones that this county wil go for fremont. . . .

Pay Particular atention to Speeking. . . . Learn to Speak Extemporanious on all and evry Subject. Strive to learn all that you can about politics and State affairs. Commit the constitution to memory. Learn all the Southern doings. Ceep all papers to refer to. Strive to be as familiar with the united States from the discovery of it by columbus as I am with my farm. Dont fear to Speak to any man let him be ever So learned. Work in your own harness. Pattern after no man. Do good to all. . . . You are on the broad Oacion. Evry term carries you fa[r]ther from shore. Learn to stear your own bark amid the howling storms and breakers. . . .

Your Afectionate Father
William H. Cook

H. B. Northup, "the same man that went south for S. Nothrop," and who spoke from the platform with Horace Greeley, was a lawyer of Sandy Hill, New York, now called Hudson Falls. S. Northup was Solomon Northup, a negro, whose father had been a servant in the family of H. B. Northup and had taken the family name as his own. Solomon was kidnapped from Saratoga Springs and sold into slavery at Washington, D.C. For twelve years he was held captive in the south but finally succeeded in getting word of his plight to his old friends in the north. H. B. Northup went to Louisiana, where Solomon was a plantation hand, and succeeded in restoring the negro to freedom. Upon his return Solomon wrote a book entitled "Twelve Years a Slave," published in 1853.

Mr. Hand, another of the local orators heard at the county fair, was a member of the Essex County family that has given several members to the bench and bar of New York state, now represented on the federal bench at New York City by Judge Learned Hand and Judge Augustus N. Hand.

FREMONT AND FREEDOM

$20 inclosed. Pleas make good use of it

Ticonderoga, Oct. 19th 1856

Dear Son:

. . . I fear dark days are comeing. There is a good deal of political feelings about here. We had a meeting up to the School house the other evening. Alfred Weed and H. B. Baldwin of Ty were the main Speekers. Weed is quite wel posted in political matters and a good Speeker. Baldwin Speeks with earnistness and feeling. Evry man in the district were presant.

Towards the close, Old uncle B. Peterson was called for. he gave us the history of the democratic party. He said they were once right but that thay had now gon astray and that he should follow them no fa[r]ther. He should go for Fremont and freedom and for freedom in cansas. He received many chears and gave good Satisfaction.

Next tuesday at 1 oclock we are to have a Fremont meeting at the Sons [of Temperance] Hall. Eminant Speekers from abroad it is said will be there. The hand bills are out. The democrats are to have a meeting soon. The knowing ones seem to think there is but two candidates that stand any chance for the Presadancy, Bucanon & Fremont. Filmore they call the Stool pigeon to draw votes from Fremont.

I am unable to keep track of these things. I want you to ceep track of all parties and all political doings both north and South. Take particular pains to instruct yourself on all Sides. Study the history of the South the doings of their eminant men the history of the introduction of Slaves into the diferant States. Work in your own harness, . . . comepare one opinion with another then judge for yourself and go fourth boldly to do good. . . .

Flave you have many great and good privaledges. I have no reason to doubt but what you wil make improvement as fast as possible. It is my highest ambition that you Should make a verry eminant Schollar. I hope you wil not disapoint me. . . . As you Seem to think that your mission is to do good here in this world of ignorance Sin and vanety it is necesary that you Should be thoroughly educated. . . . I would take a paper from both north and South of the most reliable and truthful Standing one of each party from this date thro life. . . .

Young Delano, Wickes, Kimpton are at the fort edward School. . . . J. Burnet is Stumping it for Filmore. . . .

See here boy look out how you use up money.

Good evening Sir the clock strikes eleven and I must close.

Your father
Wm. H. Cook

"The fort edward School" was the Fort Edward Collegiate Institute, then recently opened by the Rev. Dr. Joseph E. King at Fort Edward, New York. It was one of the first coeducational schools, and in later years its graduates were found throughout the United States.

SHARP RIFLES

Ticonderoga Nov. 16th 1856

Ever Dear Son:
. . . You Seam to have mutch feelings on the result of the last Election. So do we all, but I hope things may turn out better than we predict. I see no hope for poor Kansas. I dont wish to meddle with Slavery where it has a right to exist, where it is constitutional but I wil do all in my power to Stop the extention of the great evil, even to the taking up arms if that is necessary.

I want you to watch the political parties closely. By So doing you wil learn the vast Sight of deception that is practiced on all Sides, the vast Sight of iniquity that is practiced for the Sake of obtaining office. The South are true to a man for their own interests. Evry presidential election they wil begin to harp and cry about Disolving the union. This frightens many at the north and they Say that we Should treat the extention of Slavery with moral Suasion. A prety way to get read of the border ruffins in cansas. I say let the union Slide and go on to Kansas with Sharp rifles if there is no other remady. Let us have a fence around Slavery Somewhere as perminant as the rocky Mountains. But anough of this.

We lack men of pure and noble hearts that think and act alike. We done nobly as you wil Se in our own State. We all worked to a man. . . .

Ma and Loret are verry much taken up reading Dread, Mrs. Stows last book. . . . If you have not read the work pleas do

So, as it is a verry instructive book on Slavery. Mrs. Stow is a great riter. . . .

Inclosed is $30 dollars which you must make go as far as possible. . . .

Your Father
Wm. H. Cook

CANKER RASH

Ticonderoga April 5th 1857

Ever Dear Son:

. . . The Stalk have wintered wel. We have pigs and calves and Soon expect a colt from old bay. I wish I could Say as mutch for all the neighbours but it is otherwise. The Canker rash is prevailing here verry bad among children. There is but few that escapes. Its a terible disorder. Sharlot Cook has lost her youngest girl, the others barely escaping. William Cooks little boy came verry near going. I cannot Say that he is out of danger yet. Mr. Litchfield of Chilson hill has lost five by this disese within a few weeks and has one more that lays at the point of death. Charleys two children are verry sick with the disease. The oldest I think wil be no more in a Short time.

Your ma watched over to C[harley's] last night. She is now Sleeping and Says you must excuse her for not writing this time. Let little children come unto me for of souch is the kingdon of Heaven. O how consoling to the bereaved parents to think and know that they have gon home before they have committed Sin knowingly Sufficient to shut them out forever. . . .

I have always written you to try and learn all and evrything. Perhaps that is rong. You may get more into your bundle than your band wil reach round and be obliege to thro out Some in order to make a tighte bundle. You have now arrived at an age So as to understand what field you intend to work in. Therefore is it not best to use all the power on the one preperation which [?] I want you to lay a Strong and broad foundation for your education. If Providence permits we wil build it mountain high before we finish our labour.

A word as to colleges. I see that you have many invitations to attend differant colleges. You must get all the light that you can and then decide for your Self as I have no choice, only I want you to have all the advantages of the presant age. I think that you had better Spend a year or more in France Soon after you graduate to finish up your French. Launch out go boldly to your work advocate truth and temperance in all things dont fear to expose rong in what ever place you may find it high or low.

This morning as I lay upon my pillow I heard the notes of our phebe bird for the first time. She has arived to claim her home near the doar under the porch. The snow is quite gon. We are building fence on the covil farm. Shal Soon Start the plow. I have hired Abijah Bevins to work thro the Summer. Loret went home yesterday. She felt bad and Shed many tears. So goes the chequered cenes of this life Sunshine and clouds.

The colt that C[harley] and I own together is a Splendid fellow. We call him worth one thousand dollars.

Your Father

Wm. H. Cook

Pleas run this letter thro one of your screens and preserve the Good and let the bad go to the winds.

MORE OF THE LIME BARREL

Ticonderoga June 2d 1857

Ever Dear Son

As granmas letter is a going I wil write you a few lines although in a great hurry. . . . We are thro planting and Sowing. Shal wash Sheap tomorrow. The prospects are dull for the comeing clip. The fany mare and old bay have both got good colts.

We have rec^{d} all the letters and papers that you have Sent. . . . I could Say mutch as regards your course and pursuit of doing good to the worlds teeming masses, but you have council from learned men far better than I could give. See for yourself that all of your propelling powers are evenly and justly balanced and held and managed with a strong reign, So that they can be governed at all times in case of a fallior in any part of the machinery.

I wish you would be content for the future in filling up your Storehouse of knowledge for future usefullness not Spending your Strength and time in getting up Some little reformation in this or that thing. Remember that you are but a boy yet and are not sufficiently experienced in these things to reform the world.

Better Strive to imitate the lime barrel yet. Let your Lattin and greek and class studies be as the water turned upon the Seeming flinty Stone. Then put your ear close down and See if you dont hear a distant rumbling. Gather up your knowledge from all parts and dash in as you advance in life letting now and then a truthtelling Stone burst and go upward to satisfy the Sarounding multitude whats within.

Follow this course of gathering knowledge and dashing upon the now agitated movements until you arive at the age of twenty five. Then Study and practise until thirty. Then if the lime is good in your barrel flesh and blood cannot Stay your hand. . . .

Yours truly
Wm. H. Cook

LOUIS RIDES AND PREACHES

Ticonderoga, June 21St. 1857

Ever Dear Son
. . . They have all been to meeting to day but me. I Staid at home to watch the bees as they have just commenced Swarm-

ing. Lewis [Beaudry] preached in mr Wrights pulpit this afternoon. They Say he does verry wel. He has a horse and waggon now and rides and preaches all the time in Hague and Ty.

I have read over your last letter with care. There are many things in it that I like . . . , but there is that Same feeling to do miselanious business, to be in some exciting conflict to do good or to establish the truth as you say, which I think is your greatest fault.

Let others step in and do this business now, as you have done your Share. If they cant do it let it go on dun. You Should be in deeper things now. You have practised as Scrub trotter long enough. You Should now introduce yourself to older and more experienced competitors and come out on to the three mile track and try to handling the reigns there. I Se by your last letter that you pay three times the attention to other business that you do to your class studies. This I think is rong. Your class Studies are what costs you money. Other things can be attended to at less expence. . . .

Your father
Wm. H. Cook

A KITTLE UPSIDE DOWN

Ticonderoga, July 8th 1857
four o clock wensday morning

Ever Dear Son

. . . I think we have many that go fourth to do good before they are thoroughly prepaired. When they address an audience its like turning water onto a kittle turned the other Side up. It all runs of without any affect. I dont think that any man can become a perfect Orator by Study. He must get all the materials then trust to God. . . . If he is a true messenger his words will Sink deep into many a heart.

You speek of going to N. york with your friend, Stoping at yale to engage bord. Unless this rout wil be verry profitable

I think it had better be posponed until we have more funds. My money means are far below my wants to ceep you along and do my other business, although I wil send you the necessary money if you think it will pay.

I want you to have all the advantages that our means wil allow. Do take care of your health brain and head. Word came from Keesville the other day that you was crazy. I hope this may never be true although when I look at the vast amount of business you are trying to do it Sets my head whirling. . . . Ceep yourself evenly balanced, and onward for truth's sake. . . . All well her and plenty of work. . . . Shal commence haying next week. We have met with Some losses in the Sheep and cattle line Since writing to you last. Your friend Sanborn called on us the other day. He Said he wanted to see where Flave lived when at home. . . . Ma commences making cheese this morning. Says she cant wright but wil write Soon. . . . Inclosed I Send you twenty dollars. Let me know in your next how mutch you wil want to close up. . . .

Your Father
Wm. H. Cook

The report from Keeseville referred to above was a forerunner of dark days when his son had to be removed from Yale and placed in a sanitarium near Boston.

CHAPTER III

Yale in the Fifties

NOW Flavius begins his college life.

YALE BY WAY OF NEW YORK

New Haven, Conn., 139 York Street
Wednesday, Sep. 15, 1858
to Sabbath, Sep. 19, "

My Dear Parents:

Safe in Yale at last, a good room, a good roommate, a happy prospect. . . .

We left Ticonderoga at 2:30 p.m. & arrived at White Hall at 5 o'clock. . . . Driving from Whitehall to Saratoga in the cars there was much of interest to notice . . . , and so on to Albany. . . . Soon after leaving Ft. Edward, I felt a heavy slap on the shoulder & turning round saw young Mr. Kimpton of our place. He was bound for Yale, but was going down the Hudson by the night boat so that he could accompany us only to Albany Junction. He was not certain that he should remain at Yale but was going to examine with several others from Ft Edward [Collegiate Institute]. At Saratoga we had time to drink a glass of Congress Spring water in the Depot & notice the rich park in front of the principal Hotel. . . .

As to the care of baggage, freight, expenses, &c. I have added something to my knowledge on this short trip. . . . To come to the end of a busy railway route in the night, or even to land in New York in the day time, with my inexperience, amid the bustle of hackmen, porters, & hotel agents, always was a sort of terror to me. This time, after understanding the rules of traveling a little better, all was easy. Before you arrive at the end of your route, on all the principal thoroughfares of travel, you are visited by an express agent. You may safely resign your check to him, & he forwards it to any place & at any time, you name. . . . If one understands the *rules & customs of the route* it is no harder to travel right by cars & steamers than by oxteams, nor in cities than on your own farm at home. "With other things learn to travel," said father, & I have tried to do so.

In Albany we stopped at the Adams House. We did think of going to the Delavan House, which I liked from its association with the name of E. C. Delavan, the Temperance philanthropist. But on hearing the cry of the porters at the gate "Adams House, sir, only $1.25 a day," & "Delavan House, sir, $2.25," we preferred the former, especially as it was just across the way. It was 8:15 o. c. p.m. when we had finished washing the dust from our faces & sallied out in quest of a guide book down the Hudson. I found Appletons Hand Book . . . & felt surer of a profitable & pleasant trip down the Hudson on the morrow. . . .

We retired at 10:30 & were up at 4:30. The boat down the Hudson left at 7, & we were anxious to see as much of the city as possible. . . . As I looked at the Statue of Justice blindfolded on the summit of the capitol I remembered Seward's allusion to it with interest. We came back down broad State Street noticing several Banks, the rooms of the Young Mens Christian Association . . . , and turning into Broadway saw the Mercantile College, Little's Law Book store, & several other places of importance. . . .

[We had] a rainy and cloudy day on the Hudson. . . . On board the *Armenia,* the steamer on which we came, is a Calliope, or an instrument resembling an organ & played in connection with the engine. As we left Albany this band like music played *Auld Lang Syne.* It was appropriate. Then as we landed at or left a place, the Calliope was played with stirring effect. . . .

We landed at New York at 5 o'clock & took a room at Lovejoy's Hotel, this house being kept on the European Plan. You pay for your room only, & get your meals where & when you please. . . . I . . . visited the Astor House, Taylor's Dining Saloon, Brady's celebrated photograph gallery, & finally dropped in at Fowler's & Wells.

It had for a long time been an object of desire with me to have a written phrenological examination from the highest school of that system of Mental Science in the world. Fowler's & Wells, as represented by their cabinet, & the Phrenological Journal at New York, I considered the best on our continent. O. S. Fowler, for some reason, I considered the examiner of most ability and experience. I enquired for O. S. Fowler, therefore. He was not in. Would he be in before Monday? "No, sir," said Prof. Sizer, "he has not been connected with the establishment for several years." "Indeed! that is news to me. I'm an outsider. Is he on a farm?" "Yes, sir, he is working all the time he can get." "Writing any book now?" "Well, yes," said Prof. Sizer, in a tone that did not seem to indicate any very high appreciation of the work in question, "he is writing *at* a book, I suppose." I enquired for L. N. Fowler. They did not seem,—Prof. S. & his clerk—to have a very high appreciation of O. S. L. N. Fowler, they said, has not been in the office but a few weeks during the year. Prof. Sizer seemed to be the man of the establishment. I had seen him at the Cabinet a year before, & from other sources than his own assertions, knew him to be well qualified to give an examination. . . . We seemed to be acquainted with each other immediately, & discussed the Atlantic cable, a part of which I had bought & had in my hands. I told him my age, that I had received a good intellectual & moral education so far, and said on sitting down to be examined: "Now I'm just going to college, & wish you to do me all the good you can."

In dictating the examination to the young man who sat by and took it down in shorthand, Prof. Sizer seemed to be remarkably nice in his choice of words. My body, I was rejoiced to hear pronounced for the first time by a practical anatomist and physiologist, to be in sound vigorous condition, and sufficiently large to support my brain. I have never had this said to me unqualifiedly before. . . .

While speaking of my religious nature, he ascertained that I was a Congregationalist. "That is as it should be," said he, "I was afraid you were an Episcopalian."

"I saw you over at that heretic's yesterday," said he, on Monday morning, when I called for my examination that had been written out. "So you go to the heretic's too," said I. "It's the best food I can get," he replied, "I saw you swallowed every word yesterday. I belong there & have for years."

The heretic's church which Prof. Sizer mentioned, was Henry Ward Beecher's. . . .

On Monday morning Bro. Daniels & myself were up early. . . . We walked over the St. Nicholas & Metropolitan Hotels; in at Brady's Saloon & restudied again his Imperial Photographs of living senators, of Cyrus W. Field, of Peter Cooper, & other celebrated men; . . . went over the Times office; explored Barnums Museum for two hours, & finally clim[b]ed the tower of Trinity church to a height of 250 feet where we had a noble view of New York. . . .

Besides these visits to places of interest on Monday morning we went over the Tribune office. Late Saturday evening I went over & renewed your subscription to the Semi-Weekly Tribune, & was so fortunate as to find a man well acquainted with Ticonderoga to do the Business with. . . . Mr. Jenny told us when to come in to see the editors. It is "Doesticks" or Mortimer Thompson who writes the sarcastic articles in the Tribune. You know we have wondered often who it was. He was noted by the name "Doesticks" long before he was connected with the Tribune. I said that I had been told that it was Mr. Hildreth that wrote these articles. "No, indeed," said Mr. Jenny, "he is too sober minded a man for that. He is one of the finest historians, & writes political articles." A stout, jovial, leghorn-hatted man with spectacles, who was enquiring for a key to some door of the establishment was pointed out to us as A. S. Hill, the legal Editor of the Tribune. . . .

I peeped through the glass partition over the top lights which were left unglazed and saw Greeley leaning on his desk, busily talking with another man. "He is always busy," said the gentleman who showed us about, "You go in there on business & he will tell you to 'hurry up.' " . . . Mr. Greeley's Desk was the same old one that I saw a year ago, & that is pictured in his life by Paeton. It was covered with manuscripts & papers, folded & unfolded, & lying in much apparent confusion. His desk and room was much less neatly arranged than that of Mr. Dana, immediately in the rear. I saw no map or globe & very few books in any place but the library, except a large hanging map of the world in Dana's room. A pair of scisors were hung to the top of Greeley's Desk & swung down in front ready for use. An old hat with broken brim & worn nap stood on his desk full of apparently new papers & placed as though he had just taken it from his head. He was very simply dressed, & seemed fuller & portlier in figure, & in ruddier health than when I saw him for the first time a year ago. As ever, he seemed much absorbed in his business. He was moving his lips very rapidly & gesturing earnestly in conversation with a man whose back was towards me. I had a full view of the face and form of our greatest Editor. I could not look upon his countenance without reading there peace of mind, blamelessness of life. . . . He has nothing of the malicious, reckless, foxy look, of James Gordon Bennett, editor of the *Herald,* whose imperial photograph I saw and studied at Brady's. . . .

We had a noble ride over the Sound, 80 miles to New Haven. . . . We left Peck's slip at 3 p.m. . . . Just before reaching New Haven, Bro. Daniels came to me at the stern of the boat & proposed that I should be his room mate for the first term in Yale. I was overjoyed at the proposition. . . .

After a good nights rest at the Merchant's Hotel, we were fresh & happy to begin work next morning. . . . On going out to look at our room, which was about five minutes walk from college . . . , I met many old mates whose greeting was very warm and happy. Our room is a front chamber twenty feet square. . . . Our room is lighted with gas. . . . The burner juts out from the wall and elbows out. . . . We can shut the elbows back against the wall when not burning the gas. We have a wood closet and a clothes closet under the garret stairs. . . . We have a bath room & tub in the first story that we can use any day of the week for a cold bath, & can have a warm one whenever we will heat the water. All these advantages we have for . . . 75 cents a week. Other students pay from $1.25 to $3.50 for rooms no better, only nearer college. . . . I have board in a club directly across the street good enough for a king for probably $2.50 a week. $3.00 or $3.50 are usual prices.

Your Affectionate Son,
Flavius J. Cook

Brady's, where Flavius "saw and studied" the "Imperial Photographs," was a photograph gallery opened at Broadway and Fulton streets about 1842 by Mathew B. Brady, who became famous for his photographs of contemporary celebrities. O. S. Fowler, of Fowler and Wells, Flavius' favorite phrenologists, besides being the author of numerous works on phrenology, was a student of architecture. Eight-sided houses sometimes found in the northeastern states trace their origin to a work from his pen entitled "A Home for All, or the Octagon Mode of Building," published in 1854.

THE BEAR HUNT

Ticonderoga Sunday 26 Sept 1858

Ever Dear Son
. . . Last friday night the old bair paid a visit to Mr. Bisels corn field. On Saturday runners went in evry direction for

help to drive the Mountain. They assembled at Mr. Douglases at eleven o'clock about Sixty in number. The best gunners were Stationed across the mountain from Mr. Douglases to Stones bay [on Lake George], and then commenced the drift from the South end of the mountain, with dogs and guns. In about forty minuets they Started her out of a Swamp east of A. Shattucks. Awaw went dogs in full chace. She took a turn towards the [Rogers'] Slide for a few minuets and then turned north, which Soon brought her to the line of gunners which Stood in excited watchfulness ready to receive her.

They herd her jump long before She arrived. Soon She came bounding thro the brush with tremendious crash, and bang went Six guns in close Succesion which told So wel uppon the old Bear that She could go no farther. Then went up the Shouts of many voices hora we have got her, here She is. Her legs were tied together and She was brought to the clearing and then taken in a waggon to Mr. Douglases house where the parties all assembled with many others about 100 in all.

She was weighed and found to weigh 402 lbs. An old he Bair the one that has been a terror to the farmers on both Sides of the mountain. She was then Skined hung up and dressed and divided into sixty parts. The Skin Sold for $9 and divided among the sixty hunters and all returned to their homes happy for the Success of the day.

L. Covel gave her the first shot. Mr. Jones that lives with Charles Hay was the first to bring her to the ground.

Your Affectionate Father

Wm. H. Cook

Flavius had now become an author. *Home Sketches of Ticonderoga* by Flavius J. Cook appeared from the press of W. Lansing and Son, Keeseville. Numerous references to this work will be found from this stage forward, as William H. Cook grows caustic at the prudent citizens of Ticonderoga who were somewhat rueful over the price of the book, to wit, fifty cents.

MOTHER HEARS A LECTURE AND DELIVERS ONE

Ticonderoga, the last day of the year—1858

Dear Son,

Last evening your parents went to the village to hear a discourse from C[layton] Del[a]no, on Education. Called at the office found your long letter, Letter as you call it, but I do not call it so. It reads more like a lecture on Education, very much like the one that we heard. . . .

You say a great deal about taking care of your health, but I do not think you do, . . . If you did you would get on flannel. I do think it astonishing that you do not get flannel shirts, if not drawers. I feel just like scolding you. You have always worn them from a child. Now where the air must be very different coming of the sea, you go without them. But I am sure I have Said Enough. You think you know better than ma, as usual. . . .

I send you a pair of home made shoes. I made the tops, and Ann put on the bottoms. They look pretty clumsy but if you do not want them, you can put them in your trunk and fetch them home. Your Pa can wear them. . . .

Pa has gone to village this evening again, to see about sending Vinco to Illinois, as the man has come after him. . . . I will defer the rest till tomorrow.

Jan. 1, 1859. . . . They are like to have some considerable trouble about the horse. Pa will write you all about it soon, but says he is so engaged now, that he will not write. He is going to Covenant meeting this afternoon, and leave your shoes, and this letter. . . . I wish I could send you more, some of our good apples. I would like to see you and Mr. Daniels sit down and have a feast of apples from home. . . .

Your Home Sketches make quite talk here just now. They are feeling dreadfully about the price. They think they had ought to be afforded for 25 cents certain, and I guess some think they had ought be afforded for nothing, for some one undone the package, as soon as they came, and 6 are missing. Mr. Right was here yesterday. Said he was up to the north part of the town the day before, had considerable talk with a number, about them. Tryed to convince them of the expence, beside your labour. He says that they are well worth $1 dollar, to any one, and there is as much matter in them as in many dollar books. And he would not certainly sell one less than 50 cents. Your Pa thinks that they wont sell here at that price. But says it would be his way to not sell any for less here. If you are obliged to sell for less send them somewhere else. . . .

I have not much news to write you about things here for I do not go much, see much, or hear much, therefore I do not know much. I must say to you again, take care of your health and raiment, as well as the mind, for they are relation and closely united. . . .

Your ever fond Mother
M. Cook

A MIDWINTER LETTER

Ticonderoga, Jan. 10th, 1859

Ever Dear Son

As the weather is so extremely cold that its impossible to work out doars without danger of freezing, I wil Set down in our pleasant home and write you a few lines. We are all wel and happy here. Yesterday we attended the funeral of Francis Arthur, the old gentleman, at the white church. Sermon preached by Rev. Stevens [of] C[rown] point from these words, Work while the day lasts for the night comith when no man can work. He is a Sound man but Speeks quite to low for a large audience. He was burried in Masonic order. The house was full Say about 600 people. Owing to the coldness of the day mutch of the council and influence was lost. . . .

The Home Sketches are selling slow. Many find fault with

the price. All that have spoken to me about the price I have told them that they were not obliege to buy. Dr. Smith Said if they were to be had for 25 cts he would buy Some for his friends. They Seem to think that you are going to make Something on them. I know Something what they cost. Dont Sell them one penny less than you have offered them. We have plenty of men in Ticonderoga that think more of fifty cents if we are to judge by their actions than they do of their own Soul. . . .

Mr. Howard gave a lecture on reading the other evening. Some parts of it was verry good. The house was full. He proposes to read Shakespear [in the Academy] a few evenings if the trustees wil permit having pay taken at the doar for admision for Say 6d. I shall oppose it, as any thing of a theatrical performance wil Start out all the drugs of Ty equal to a horse trot or a dog fight. . . .

Your father,
Wm. H. Cook

Flave, have you put on your Shirts with Marsailles bosoms and long wristbands? If not you must, and wear them through cold weather. And I want you to be sure and tell me in your next weather I shall make any more wristbands in that way as I am about making you some more. Buy you some sleeve buttons. You have seven of them shirts. Lay your others aside and wear them. They will be warmer, and do better for winter than summer.

All love
Mother

HOME SKETCHES STILL TOO HIGH

Ticonderoga Jan. 30th 1859

Ever Dear Son

. . . I am drawing rails Some, getting boards and Shingles to build and repair the barns in the Spring. The thrashing is not done yet. Charley and Mr. Brown are busy thrashing every day. . . . The grain is yealding well. The Stalk is doing well. The Sleighing is good here. The Home Sketches are Selling Slow. . . . I think we will work them off So as to pay.

I want you to concider where you would of been if you had of been depending on the Sale of the home Sketches for your Support. . . . G. D. Clark thought they were quite too high. Said he wanted to buy Several for his friends. When you talk about Liberal men in Ticonderoga I would like to See them in a cause of this kind. The Home Sketches I think are well arranged and . . . wil be the means of doing good yet, and mutch Saught after. Dont use them carlessly. You are now of Age and your own man. May your future be as happy as your past life. Let us Pray God when in prosperity to prepair us for all the changing Scens of life. Do all you can to help carry yourself thro college, if in no other weigh than by being prudent and Saving with your meens.

Your Father
W. H. Cook

MOTHER'S THRENODY

Dear Son,

. . . Your Pa cautions you very particular in every letter to be prudent and saveing. This is all right and well. But I can tell you that you could caution him in the same way and I think there is more need of it, as everything is his, and he can do as he pleases. He does just as everybody want him to do, (except me). If a man wants a dollar, 5–10, or even 20, he can have it, and he never makes a trade without throwing in something, sometimes a good deal, and it does seem to me as if there had been nothing paid for 2 years past, for it is pay debts and pay debts that I supposed was paid long ago. . . . I have felt anxious for him to put the $3000 that he made on the sale of that farm at interest, and save it for you till you get ready to commence business. This he is now trying to do, but you and he will have to be more prudent, and not give a way so much, or he will have to use it, before you get through College, and I should not wonder that if before 10 years come round, this farm had to be sold, if things go on in the same course that they have been for two years past.

You will say Ma is in a real fret. Well, I did not intend to write what I have, when I begun, but some way or other it came along No, I am not in a fret, for I have made up my mind to it. . . . I have made up my mind that my anticipations that I used to cherish when we were getting propperty are vanished. The extras that I used to think and was working for, when we got able I see I cannot have, or have willingly (for Pa very often says he had ruther give it away than to have me have it,) so I will be content and never visit my friends, or have any extra luxuries, if I can only please him. . . .

O, what would I give to see you, how I wish you could be at home.

Your Mother

ALL HAPPY AS SPRING COMES

Ticonderoga March 12th 10 Oclock pm, 59

Ever Dear Son

As the rain is falling fast and the flatts begin to flow I Set me down to write to you. Ma has just layed down to rest her. . . . Anna Sits by the table in the kitchen Sewing, the kitties lay on the Sopha pering, the boys are Sawing and Splitting wood under the Shead and Singing and whistling. All seem to be happy. God desighned us all to be happy, but thro ignorance, Sinfulness, petulence, anger, Strife, evil doing and many other evils, we cause ourselves mutch misery. . . .

I think we can dispose of Some of your Sketches this Summer to travlers. We must make them pay Some how. We have received the F. Leslie Magazine and paper likewise the paper containing the description of H. W. Beacher, for which we return you many thanks. . . .

Elder Wright has held 3 meetings up at our School house the past week. Tolerable good attendance, but no comeing forward to duty, weighting for mor light. Prey God that this light may be given to all. Clayton gave us a good lecture on agriculture. A thin house 150, or more, mostly young men and ladies. I have Sold the old Oxen for $120. . . .

Your Father

W. H. Cook

Hardly had Flavius arrived at Yale when he was inspired to found an international magazine, to be published by undergraduates of the leading colleges and universities of the United States and Europe. The publication was finally under way and was called *The Undergraduate.* The name was later changed to *The University Quarterly.* After Flavius left Yale in 1860 the *Quarterly* did not long survive.

FLAVIUS FOUNDS A MAGAZINE

Yale College, May 29, 1859

My Dear Parents:

For several weeks I have hardly had time even to write home. The printing of a new circular for the *Undergraduate,* my regular studies, composition writing, several class meetings, the Prize Debates, general reading, exercise, social intercourse, have kept me profitably and delightfully busy. . . .

Mr. Chamberlain pays a part of the expence of printing the Circular, & I think that I shall not be obliged to call upon you for any *extra* funds in order to pay my share or to pay the Prizes. Those Prizes are really offered in my name, though nominally in the name of a friend of *The Undergraduate,* not connected with any college. Who that friend is no one knows but myself, and I only know as I trust to my own ability to earn seventy-five dollars or to save a part of that amount & earn a part between now and such time as the prizes may become due. . . .

I have consulted the most judicious men on our Boards & they all agree that $75 could not be better spent. . . . I was not myself fully convinced of the moral expediency of offering these prizes, though I was the first to suggest the idea. . . . Because the friends of the Magazine were almost unanimous in favor of the offer, I overcame my scruples far enough not to oppose what a majority thought best. . . . I detest college prizes & deplore their moral effects as much as ever. . . .

In class meetings lately something of College life & of the character of our class has been displayed. When Freshmen become Sophomores, as we shall in about three weeks, it has been usual to celebrate the occasion by a nocturnal & often riotous gathering called a *Pow Wow.* The class get together, usually in front of the State house steps, most of them masked, & tooting tin horns, or shouting noisy College songs. Speeches are delivered which have often contained smutty allusions & abuse of the Faculty. After the speeches a procession is formed & the class, blowing horns, cheering, rattling tin pans, & headed by a band, march about the city, seranade the female seminaries, and come home toward morning, all of them monkeys, & some of them drunk. This is a fair definition of Pow Wow as I have finally learned it. . . .

At the first meeting of the class concerning it, being called upon, I expressed the opinion that it was opposed to College law, in at least three points, the masking, the night assembly, & the disorderly procession. Leave these out, & perhaps the Jubilee over being Sophomores was well enough. Nothing could have been more unpopular than to quote the College laws in an assembly of Students bent on breaking them. Chamberlain immediately rose & said "that the Pow Wow was undoubtedly contrary to College law but that we broke many college laws with a good conscience, & that we must break the college laws now out & out, and have a Pow Wow."

This was the popular side, & his motion to have a Pow Wow was carried *nem. con.* though many did not vote.

Your son
Flavius J. Cook

PA AND HIS POCKETBOOK

Ticonderoga June 3[d] 59

Dear Son

In the midst of hurry I Set down to write you a few lines. We are in the midst of building barn Shead, and painting the house. 8 hands all the while. Expect to raise the Shead frame this afternoon.

All wel and Striving to do our duty. Recd your long letter last night with mutch pleasure. No time to coment on it. Sorry to hear that you are out of money So Soon. Should the expences of this year continue I think that I could Stand up about three year. My income is Small. I am willing to do all that I can, deprive my Self of many comeforts to do good to or for others. We have got to loose on the Home Sketches. We have no money to Spend on experaments. You can do as you please about Selling the things you mention in your last letter. I tried to make you believe your trunk was to expensive, but no pa was not right. I am called on by the church to help largely in repairing church, build fence, make roads, other denominations want help, the Acadama must receive my aid, the temperance cause must receive a lift now and then, Mas parlour must be furnished, painting fixing and building must be done, the calls from Yale must be attended to, my farming help must be paid, our help in the house must be paid. Now if you can tell where the meens for to Supply all these wants are coming from with no Source but my little farm you can do better than I can. Pray God to direct.

My asociates at the excise board are verry clever fellows. . . . [But] no licenses in our place as yet. . . .

I am glad of your Success in the undergraduate. I think its to [heavy a] load for one so young to carry thro and uncalled for at the present time. Its one thing to print a book and another to get men to read them. The class of community that needs light read but Little. I look uppon H. W. B. Sermons just as you do. They are food for poor me. . . .

Your Father,
Wm. H. Cook

Friday afternoon, June 3

Dear Son: . . . Your Pa has the horrors some as we might Expect after furnishing the parlor, and is now in the midst of building. . . . You and Pa appear to think, that [I] shall be perfectly happy, and my wants will be all supplied. To be sure, it has ever been my ambition, to have a good house, and well furnished, but I never thought it would make me happy. It is one desire accomplished, one object obtained, and our Nature is such that there is no End to wants: Every day brings forth its wants, some new object to be obtained. We are never Satisfied: Or, at least I am one of that sort, and I do not know who is not.

Yes: I feel quite pleased about the furnising of the said room. I have got such furniture as I wanted, good as I wanted, and not so expensive as I expected, (what I got cost about $250.) but I expect all the enjoyment I shall have will be alone, as you and your pa do not care for any such things. You both would rather have given the money away. Pa has already said as much. But he cannot have that money: They will have to wait till Some more is made.

I had a fine time and a good visit at Whitehall, and Cohoes. Halsey and Aunt Meranda went with me to Albany. There we found Mr. Frazier, who assisted me some. I was not in the city, but about 3 hours, so you may judge what time I had for observation, beside doing my business. I saw considerable at Cohoes. That is quite a manifactureing place. I went into a cotton factory, a knitting Do. and a pin factory. And that was more than I ever saw before. . . .

Well: now I am going to talk to you a little, and I think you deserve a good scolding, yes a good scolding, and I feel as if I would almost whip you if you were here. A child of mine, all the one I have in the world, and on whom I would be glad to lavish every comfort, have supplied him with over $400, and he sleeps on a straw bed, without a pillow to lay his head on. What enjoyment do you think there is for me, with all the comforts I have about me, when I think how my only child fares. But go it, if this is your choice. . . .

If you had rather lay out your money for the benefit of others, and deny yourself of the actual comforts of life, . . . why you have no one to blame, but yourself. But I think you had ought to wait till you Earn your own money and are less expence to your pa. I do not wonder that he feels discouraged. . . . He is willing to do all that he possible can, for you and others, but cannot do for every thing that is wanted. The old adage is "that you can ride a free horse to death." I think it may possible be so in Pa's case: For every one is after him, (or his money) and every dollar goes as soon as got, and I guess some faster. But he does [not?] say much only when you and I want. He has Said considerable today, about your living as you do and working so hard, and spending so much. . . .

You say that you will sell your watch to get money to help pay expences. . . . Now I say do not sell your watch or trunk, for I cannot see as there is any policy in it, as you would only have to have another. Now we send you $20, and I want you to buy a pillow and send me word in particular that you have got one. . . .

Ma

Pa has not told you that Fanny has got a colt that he asks $600 for. Has been offered $300 for him. . . .

MUSIC IN THE AIR

Ticonderoga june 16th [and 17th] 59

Ever Dear Son

I have just returned from meeting. Have attended three Sermons today but alas how few take to heart the Solam truths of the bible. . . . Ma is a little out of Sorts this evening, for She has just rec^{d} a line from aunt Maranda Grant that She and Hellen and Halseys wife and children are to be here to-morrow. We are new paint everywhere. O what a time. . . .

I have had pretty hard work to prevent licences being granted in Ticonderoga. Evry licence has been granted that has been aplide for except in Old Ty. Our bord I think are quite to willing to grant licence without knowing the principles of the man applying.

We are verry busy. The music of Saws, the cutting of plains, the rubbing of brushes, the clipping of Shears, the clinking of hows, the rattling of plank is to be herd about our home in the valley from the rising of the Sun to the going down there of. I have had 12 hands imploid today all told in various kinds of business.

I leave tomorrow for E town on the noon boat to Stand errect for the temperance cause in old ty. I work Still but Sure without a Shaddow of turning, let what wil come. I am after them Sure. . . .

Your Father
Wm. H. Cook

MORE RUSTIC AIRS

Ticonderoga July 17th 1859

Ever Dear Son

. . . The hammer has ceased its nois except now and then a Sound from one man as the doars are being hung to the new barn. The wheating of Syths, the rattle of the mowing masheen, and the Jolly laugh of the Stalwert man as he wipes the Sweat from his brow, as the Summer is verry hot here and dry, may be herd now in our valley. . . .

When I look at the expences that I have been called to bear the past year in carrying you thro college term, in building, in fixing and furnishing the house, in paying help both out of doors and in the house, in doing my farming, . . . and in the many deeds of charity and for Society I cannot tell how I have come out of them all unless the arm that is over all and under all has held me up and kept me Safe thus far and to him be all the praise. . . .

Your Father
Wm. H. Cook

TUCKED UP FOR WINTER

Ticonderoga Nov. 6th 1859

Ever Dear Son

As ma has writen to you today I presume She has told you all the news. She has probably found Some fault with me. I have nothing to Say in my own behalf. May her eyes be opened to See the blessings that God has bestowed uppon us all, and may her heart be truly given to God. . . .

The old exchange on the corner is being fitted up into a fine country store by Treadway & Weed. The appearance of our village is improving verry mutch. . . .

The gethering in of crops is over. We have plenty to Satisfy the wants of the comeing year and Some to spare for our brother who Stands in need. We are comfortably tucked up for the winter, yet anough to do to fulfill the command to be diligent in business. We have plenty of good reeding. Wendal phillips address at N york is bold Spokin. H. Wards Sermon on O Brown doings at Harpers ferry begets new thoughts.

And now and then a paper from you, whitch is thankfully rec[d]. . . .

Your father
Wm. H. Cook

YOU AND MA AND ME

($30, inclosed) Ticonderoga Nov 20th 1859

Ever Dear Son

. . . Next tuesday if you Should hark you may hear Some Squeeling as we intend to kill our hogs on that day. . . . I have Sold my Sheep off to 210, in all 11 head of cattle 9 Horse kind. Plenty of hay to keep this Stalk. The horses are all right. Getting Some of them into the Stable. Our Shead is chuck full of dry wood, our Sellar well stored with roots and apples and our cup Seems full to overflowing with the gifts of God. . . .

I did not know but you would take a trip to Washington this vacation. I would get out of the citty and get an airing Somewhere. . . .

Ma Sais you had better use the five dollars over and above what you called for, to buy you a pair of pants for cold weather. Ma is for clothing, I for laying up for a wet day, and you for books and knowledge. Let us all work together. . . .

Your Father
Wm. H. Cook

A HORSE TRADE AND THE WILL OF GOD

Ticonderoga, May 20, 1860

Dear Son

. . . I hope you will not get into any other enterprise that will take your mind and attention from your Studies. Strive to become a good Schollar and Stand high in your class. This will give you power in future life. Strive to make yourself easy in Society of both Sexes if conveniant. By So doing you may bring out gems from the hiden oar within.

You Say you allmost doubt the Bible at times. I think the reeding of H. W. Beechers Sermons would be food for you as they are for me. Let us obey our better judgements and do right in all things. We all lack faith, we all commit Sin constantly by quenching the influence of the holy Spirit. May God pardon us all for the Sin of unbelief. . . .

Old fanny folded last friday. Has got a fine bay mare colt. I have Sold her last years colt that is a yearling now to Biron Woodard of Saratoga for eight hundred dollars. He paid one hundred dollars down and pays a hundred dollars a month untill he is paid for. He is called the finest colt in the county. If this traid goes off according to agreement and I get my pay according to promice I shall look uppon it as an ordering of God. . . .

Your father
W. H. Cook

STERN WORDS FROM FATHER

Ticonderoga June 17th 1860

Dear Son

. . . I and Ma have been out to church today. There has been an effort to Start our Sabbath School but alas how few are willing to lend a helping hand in the cause of reform. . . .

Clayton has Sent up a list of names to the District atterny to be Sopeenied before the grand jury at the june term. I cannot Say what the result will be, but I think about as usual. We have no whole hearted temperance men in old ty. . . .

And now I must Speek to you about that that causes me mutch Sadness. When you first left home for College I felt that I had a Son that could carry himself thro College and Stand errect ammidst temptation paying his own bills as Soon as due, Showing to the faculty and Students that you knew the worth of money and time, but alas we are all doomed to disapointment. I have allways thought that there was a cloud over your letters or that there was Some covering up or ceeping back of the true condition of all your affairs. I have asked many leeding questions to draw from you your true Standing in financial affairs but to no purpose. Now this dark weigh of working I dispise. I Send inclosed letters and term bills that I have forwarded money to pay. I cannot make this bill corespond with the catalogue. See to it that all is right and preserve all these papers that I Send. If you owe money to your wash wooman and Seemstress just tell them to Send up their bills to one who believes that part of his duty to God lies in paying his debts. . . .

Your Father
Wm. H. Cook

Clouds were now gathering over the Cook household. Joseph was beginning to show symptoms of a deranged mind, brought on apparently by morbid religious broodings. This malady finally resulted in his leaving Yale.

WIDEAWAKES ON PARADE

Ticonderoga, September 16th 1860

Ever Dear Son

Your two last letters are received and have been read with pleasure, but when we look back uppon the last days that you Spent with us and See how you were carried about by the different moods, we fear even before the reception of those letters that a cloud might of passed over your prospects and that you are not able to controll your feelings. These things give us Sad feelings and thoughts. I am quite humbled in my estimation of your powers, Since you have been thrown from your ballance by the acquaintance of Miss H.

You have plaid the butterfly in good earnest by flying into

the candle and Scorching your own wings and falling powerless to the ground, and all this in opposition to your parents advice and your better judgement. . . .

Tuesday found us, Ma and myself, on the way to the fare to Burlington. The next day was rainy which made it unpleasant. We had quite a good time in the afternoon on the ground. Thursday morn pleasant, went up round the College grounds and thro the main Streets with proffit and pleasure. At ten I took the boat to go to Plattsburg to a great Rep[ublican] mass meeting, and Ma and J. Cooks wife are to return home today and I am to follow this evening. [This letter though dated at Ticonderoga was apparently started at Plattsburg and continued upon arrival home.]

We collected on the fare ground at Plattsburg there to hear Speeking from different persons, Stanton, Chittenden and others. Estimated six thousand presant on the ground. In the evening we had Speeking in the court house and a large company of Wideawakes with lighted torches and fire works. At ½ past ten left for home. Arrived at the Old Fort at 6 in morn. Walked home and found all things right. Friday finished cutting up the corn. . . .

I will Send you Some money as Soon as I can get it. Don't brood over the past and get gloomy. Strive to ceep clear of trouble with your female acquaintance. . . .

Your Father
Wm. H. Cook

On considering the matter I conclude to send you $20. . . .

More of "Miss H." anon.

A. LINCOLN'S ELECTION PREDICTED

Ticonderoga Oct 21st 60

Ever Dear Son

. . . Our town fair went of verry respectable. The avails of the grounds were Some $350, nearly anough to pay all expences. A band from Glens falls and the invincables gave good Satisfaction. . . . All honor to C[layton] Delano for carrying it thro. . . .

We have a company of wideawakes in town of about fifty. They went to Hague the other evening and had a good time. Had an address at the meeting house. There was a large number preasant. I think that A. Lincoln will be our next presadant.

The council met for the purpose of ordaining B. Ashton last wednesday but could not agree and So ajourned to meet in Manchester the Second tuesday in Nov. there to meet his accusors as they refuse to come here to ajust all difficulties. I may go. . . . As I am called to mingle with proffsional men of high Standing I See a great lack of the right Spirit of love one to another. . . .

I give you leave to take what course you think best as regards the prizes for the best writen piece in the Quarterly. If I have got to raise the money, give me long notice. . . .

As regards comeing home to vote I hardely think it will pay. We Shall do our duty on the 6 Nov. . . .

Your father
Wm. H. Cook

TRIAL OF A CLERGYMAN

Ticonderoga Nov 18th 1860

Ever Dear Son

. . . The letter you wrote me directed to Manchester, I did not receive. I will now give you a brief discription of our doings. Last Monday eve found Messrs. Rev. Bigelow of Keeseville, Burwell of Moriah, Cook, Bly & Ramsey taking refreshments at the America House, Burlington. Immediately after we left for Manchester in a Sleeping train their being no change of the Sleeping train from here to Manchester. Thro fare $6.70cts. There was a good deal of fear manifested by

Some that there might be axidents. I trust all Souch things to God feeling as Safe as I do in my own house. Many pleasant thoughts and a comfortable ride during the night, but little Sleep.

The next morning at early dawn found us at the depo at Manchester. Our friends Grant & Foster came down in the day time monday. After refreshment we were met by Rev. Ashton & Foster. Soon repaired to a quiet place belonging to the 2[d] Babtist church and called the council to order, Rev. Burwell taking the chair. . . . We are here to here evidence as regards the ordaining of Rev. Ashton not in judgement against the Manchester church. After Prayer by Grant we Sent for the church to present their grievances. Their acusations were, Slander, . . . , falswood and the obtaining of the Signiture of a lady by compulsion. Their proof of those charges was verry weak in all the acusitions except falshood. This they Proovėd by Showing that he once, soon after his comeing to Manchester, promiced to marry a young lady. I was one of the committy to visit the Lady and father. They both testified to the promice, Ashton asking the consent of the father. I will not go into the details. Ashton made his defence quite humble asking pardon of the Bretheren of the Manchester church for all offences. It was not granted becaus they Said they did not believe that Ashton was a Christian. Here the trial ended and the council voted that it was not expedient to ordain him at preasant.

The council from the Ty church took the liberty to ask Br. Ashton to go home with us and labour with the Ty church. In time, if he conducts himself well he will be ordained. He has faults and who has not. I think him a persicuted man and may God give him Strength. . . .

Your father
Wm. H. Cook

While Flavius was a prey to religious ravings his condition became aggravated by his falling in love. The young lady was Miss Georgia Hemingway of Fair Haven,

Connecticut, and their meeting at this time was the beginning of a romance that led to the altar seventeen years later.

NO MORE MILK AND WATER

Ticonderoga Dec 14th 1860

Ever Dear Son,

Yours of the 10th came to hand today. I must Say that I am perfectly disgusted and out of patience with your manner of writting, and complaining of trouble and Strugles. You are now reaping the reward of doing So mutch misilanious business heretofore. You find that your mind has been overtaxed and your faculties impaired and now when you want to urge your team to a faster pace or to move a greater Load you find them kneeling or laying down in the furrow, and if I were to judge from your letters, I Should think that you had commited Some unpardonable crime. If all this fuss worry and Struggle comes from a love affair with a young girl, then I must Say that I have been well paid for the pains and money and care expended uppon you, expecting a Strong man at twenty three but instead of that have got a granny that cannot bear the first cloud that is brought over him by his own acts.

The weigh to get out of trouble is not to get into it. If your faculties are becoming impaired So that you cannot takare of your Self, you had better advise with Some phisition, and come home where you will not be quite So mutch expence to your parents. I cannot furnish means to be Squandered in this weigh. You must do better or you must be thrown onto your own resources. I mean this in deep earnestness. We have talked milk and water long anough.

I think it verry unjenerous in you to call on me for the prize money, but you Seem perfectly willing to ride a free horse to death. . . . I fear that this verry effort [for the *Quar-*

terly] costs you your life or that that is worse, an impaired mind. I shal forward you by check $150. If you are not capable of paying up your bills and keeping all things Straight I wish you would get Some of the faculty to do it for you, as it is hard to get money under the presant crisis. I want you to render an account for every dolla. . . .

Wm. H. Cook

FLAVIUS IN THE ROLE OF SAMSON

$5.00 inclosed — Ticonderoga Jan 20th 1861

Dear Son

Yours of the 10th is at hand and I have read it over with care. All I can make out of it is that you have laid your head in Some Delias lap and have been Shorn of your Strength. You will have to wait until your locks grow again before you will attain your former Strength.

In my last I advised you if you could not take hold of your Studies with more cheerfulness and proffit, that you had better go into business of Some kind or come home. I See by your last that [you] choose to do otherwise. . . . I have come to the conclusion that I Shall be unable to grow the man out of you that I have allways intended to by the help of God.

My advice to you is to round off your Education as Soon as possible and go into some business that you can earn your temporal wants. If your Sky is So loured that you cannot go thro now you had better gather up your little all and come home for the preasant even if it makes a great talk. If you cannot find peace in the promices of God you cannot find them any where. You Seem to look for peace out of your own efforts. Step out. You can walk uppon the water with full faith.

Its hard times for money. Be prudent. As regards the presant crisis there are as many opinions as men. I don't believe there will be much blood Shed. It's a Sort of a Scare effort by the

South. . . . Mr. Sewards Speech is Spoken highly of by the Democrats. I think its verry conservative Saying nothing of the rong acts of the South. I expect a compromice of Some kind. I Say Stand firm, carry out the Federal laws Strictly and let the traitors Suffer. They have always Shorn our Samsons of their locks and I fear they will do it now. If they Slip their head out of the halter now it will take a Rorry to tame them. So mutch on seccesion.

Our nation Should be ruled by the majority if not peaceably then forcebly.

And now about your expences. You Speek of Spending $50 in your Washington trip, $50 for your prize money and what has become of the other $50? Its pleasant to know where the money that we have earned by hard toil goes to. . . . You will never know the worth of money until you have to earn it your Self. I want you to think of those things. The day is coming when you will have to paddle your own canoe. You Should practise Some now. You have lost the love for your lessons by doing So mutch miselanious business. I cannot think but God is prepairing you by affliction for Some future usefullness. . . .

Wm. H. Cook

CHAPTER IV

War Times

THE CALAMITY foreshadowed for months came early in 1861. Flavius was removed from Yale, suffering a mental ailment. All his father's hopes apparently were blasted. The son who had been the object of his anxiety and care all these years and whose education he had made the end and aim of all his labor was taken to an asylum at Somerville, Massachusetts. He was under the care of physicians for two years. At the end of that time it was decided that he might return to college under strict orders to avoid excitement and controversy. But Flavius did not return to Yale. We find him once again aboard a Lake Champlain steamer, this time bound for Harvard.

MEETING WITH NOAH PORTER

Cambridge, Sep. 2, 1863
Harvard University

My Dear Parents,

Chase shed tears for the first quarter of a mile after leaving Corvallis, whether for what he left behind or for that to which he was going as a prospective soldier, I hardly knew, perhaps for both. On the boat, I saw him drinking some lemonade, but he apologized warmly for being seen near the bar & received with evident gratitude & attention the repetition which

I made of advice which Mr. Bond had already given to him to keep out of bad company, & to learn to say 'No.' . . .

On the boat in the upper saloon one of the first faces that attracted my attention was that of Professor Noah Porter, "Professor of Moral Philosophy & Metaphysics & Instructor in Didactic Theology" at Yale. . . . We fell immediately into a conversation concerning the management of the insane, a topic brought up by my mentioning my illness. . . . He made several inquiries concerning the Mc Lean Asylum, & thought the absence of a chapel & of daily prayers, a great religious and remedial defect. He had traveled in Canada about a year before, I understood him to say, & was now just returning from there. The Canadians were "much more moderate" in their remarks concerning our war in his last visit than they had been on the former. Canada was destined, he thought, to be independent of England, but not soon. . . . The tendencies toward anexation were very weak just now: the Canadians were afraid of our heavy coming & present debt. . . . He thought that all that could be said concerning the best students of this Institution [Harvard] being more finished than the best of Yale, was that the best students here had usually better social influences in their own homes: the family education of many of those from Boston & vicinity was superior. . . . He presumed that the Vassar Female College, to be opened next year at Poughkeepsie, would be a superior Institution. . . .

Your only son,
Flavius J. Cook

A MOTHER'S ANXIETY

[Ticonderoga, September 14, 1863]
Tewsday Eve 30 m. past 8 oclock

. . . I am afraid that you have got a wrong boarding place, one that will lead you on to those exciting debates, that will occopy your time and mind. Beside I am afraid that you are not able to endure it. *Do, Do, Do* let every thing drop, but your studies, and take care of self for two years to come. Then if you can reform others and make things better, do it. But at presant take care of *self*. Think of the expence, and what would be your chances, should you fail to go through college this time. Now *do,* I must say it again, do keep your self quiet and easy. Pa say that he does not care for the expence providing you come out all right. . . .

Your Ever fond Mother
M. Cook

A THORN IN THE SIDE

Ticonderoga Oct 21st 1863

My Dear Son:

Your ambrotipe and letter of the 14th have been Recd. . . . Finished haying the 13th Oct. . . . The corn and potatoes are nerely Secure. Mr. Bond Started the plow today. I Sold my Wool today for 65 cts pr. lb. Likewise 20 of my best ewe Lambs for $17.50 ct. pr head.

We had an interesting and proffitable Lecture the other Evening the 15th from a man from Texas the Hon Mr. Sherwood on the cause of the rebellion.

Our course of reading is not regular yet. We have mutch to do. As the evenings increase the farm work closed up, we will try to addopt your plan of reeding. Barn[e]s notes, the Bible and Religious Insiclopedia are my principle reding. . . .

Dr. Bronson left about the time you did as Sergion in Grants division I think. . . .

Everyone Seems unusially bussy Striving for perishable things. May providance give them light.

I have a thorn in my Side. You are paying to mutch for Board by a half dollar a Week, Surę. I hope you will make $500, pay the expences of each year at Harvard. But you will

Say these are groveling thoughts. It may Seem So to you but not to me who has to do the furnishing. . . .

Your father
William H. Cook

DRAFT, BOUNTY, AND TAXES

Ticonderoga Dec 21st 1863

My Dear Son,
. . . I have been alone Since Bond left. . . . [He] talks of Enlisting. There is great excitement here on account of the comeing Draft. The Town have offered $500, bounty to volenteers. This town has Some 38 to raise. There is a prospect of getting the volenteers if we can get the money. It will make about $500, dollars difference in my tax if they rais money enough to get the 38 men. I do not believe it is right to pay Souch bounties. Your name is on the roll as one liable to be drafted. Be a little prepaired for the call. . . .

The future looks dark. This war is bringing us all on a level as regards temporal things. There is Something rong Some where. . . .

Your father
Wm. H Cook

AT THE STATE LIBRARY

Harvard University, Cambridge, Ms.
Wednesday Eve., Mar. 2, 1864

My Dear Parents,
. . . Dr. Bronson & myself were the only passengers on the stage all the way from Ticonderoga to Whitehall. He was on his way to Fortress Monroe. When at Vicksburgh, out of 16 negroes he examined for the purpose of determining whether they were fit to be volunteers, only 4 were without welts,

bruises & scars on their backs & limbs that unfitted them for service, some of the wounds being those of bowie knives & others evidently of burns. . . .

At Benson the cows were lying in the sun on the ashes of what was once Mr. Ladd's Hotel. We took dinner in the chamber of his store, & Mr. Ladd afterwards entertained us below by some rather Copperheadish comments on the war. . . .

The ice had occasionally an air hole, the driver said, but I felt entire confidence after being on it ten rods. It was probably nearly twenty inches thick. We started from Ticonderoga by my watch, at 11, & arrived at Whitehall at 5, stopping one hour at Benson for dinner. . . .

[In Albany] finding the Adams House, that I had intended to stop at, shut up, I went to the Delevan; on Sunday evening to Stanwix Hall; & on Monday Evening to the Dunlap House where Dr. Bronson was, the board at the Delevan being $3.00 & at the Dunlap $1.50 as I paid, though I believe $2. is the price for such a room as I had. Dr. Bronson being a soldier, & I staying over two nights made the difference. . . .

Monday Morning as early as the State Library was open . . . I was at the work of investigating anything it contained concerning Ticonderoga & not published in the Documentary & Colonial History. I introduced myself . . . as a student of Harvard & mentioning Judge Burnetts recommendation I was with kindness enough put on the track of the catalogues, a second Librarian who appeared to be the author of them directing me at first. The latter also referred me to a book on Ticonderoga which had lately been published 'by some one who lives there' & on my stating that I was the author of the pamphlet & asking casually if there was a copy in the library he brought to me the Home Sketches capitally bound.

I used all my time till five o'clock on Monday in the Library, & on Tuesday all my time, or nearly all, in the same way, & had just enough to finish taking notes. . . . I saw several maps of high interest that I had never seen before, of the Lakes & the Fort, &c. I found no evidence of a road from Howe's Cove down the west side of the outlet of Lake George.

But a map, representing the country in 1759, shows a road the whole length of Lake George on the West Side, & running through Ticonderoga by the way of Trout Brook Valley, the Wicker Brook, & the Vineyard road, or thereabouts, & so on to Fort Crown Point. . . .

I saw the original papers describing the surveys of the tracts granted in our town to Stoughton, Kennedy, & Kellet & also a copy in the Secretary of State's Office of the grant to Stoughton. The 25th of July is the correct date of this latter paper & so of the Centennial. [F.J.C. was collecting material for his address at the Ticonderoga Centennial, July 25, 1864.]

. . . On Tuesday Eve. after my investigations among the papers was over, I had about two hours to wander through the vast variety shop called the Albany Army Relief Bazaar. . . . I had the fortune to see Palmer's marbles & Palmer, the sculptor himself, & to talk a moment with him, he pointing out two of Church's Studies in the Adirondacks, one of which, a morning view of Mt. White Face, over Lake Placid, interested me much. . . .

Your aff. son
F. J. Cook

FARM AND WAR NEWS

Ticonderoga March 15, 64

My Dear Son,

Your two letters have been Rec[d]. We thank God that you have been permitted to procede in your Studies again. Ro[w]ena is with us now. We are Sitting arround the table just as we used to do before you left. Ma is reading in the Bible. All well. The rotene of chores is the Same as before you left. There is Some young lambs comeing along. Lions has returned the colt. I have turned her out. There has been a fall of two feet of Snow Since you left. There is plenty of mud here now. . . .

Town meeting resulted in the Election of the Republican Officers. The vote for the Soldiers was all in favour of the Soldiers voting at our next Election. There has [been] Several of our Soldiers from the 96 returned on a furlow. They look well. Densmore, Walker, Rich are among them. Johnny was with them but lost his money and pass and had to return to Albany. I am looking for him every day. I have received a letter from Rev. Bond. He was at Stevensburg, Va. when he wrote. I think he was in the K.P. raid. . . .

Ma and myself was out to Jacksons [Cook] the other day. . . . One of Jackson's boys found the other day a cup on the shore of the lake with a handle to it marked L. P. on the bottom with a crown over the letters. I think the cup is pure Silver about the Size of a large tumbler.

Your father
Wm. H. Cook

MORE OF THE SAME

Ticonderoga May 11th 1864

My Dear Son

. . . We have just been reeding the sucess of our Armies with joy. And return thanks to God for victory. Francis, your mother and myself are Sitting arround the table reeding the papers and Bible. Ro[w]ena is washing potatoes. It is raining this evening. It is verry wet here this Spring, So mutch So that we cannot progress with farming to mutch advantage.

The colt that broak his leg is Some better. He can get up and down alone, is out to grass. . . .

I saw Kimpton the other day. He looks healthy. He Said he Should be away at the time of the delivery of the Sentinial which he regretted verry mutch. . . .

Business is lively at the village. I think that the Graffite company will be the means of great improovement in Ty. . . .

Your father
Wm. H. Cook

LINCOLN CARRIES TI

Ticonderoga Nov 9th 1864

My Dr Son

It is nine O clock and they have all retired for the night. I have been making out my Town and county Act this evening and have onely time to write a few lines. All well.

Its verry difficult to get help to do work on the house. Have to pay $4 pr. day for a man to plaster. . . .

138 Majority in our Town for Lincoln. Clayton [De Lano] and Hiram [Kimpton] were the Speekers at a Political meeting Last monday Evening. They done finely. Col. Calkins Jr. made Some good remarks. . . .

I have had no time to consider the Oil well Speculation. I am attending to Black Wrincle and from this Source expect Some proffit.

I am Glad that you think occasionally about doing Something for your own Support for I am on the downward Train and cannot always Supply your wants.

O. Walker that went into the Servis reported dead. Mr. Roberts Son wounded. The rest of the friends all wel as far as herd from. O this terible war. May Providence Guide us as a nation and may we be Submissive in his hands. I fear for the result. We are a wicked, fast people. . . .

Your father
Wm. H. Cook

THE FALL OF RICHMOND

Ticonderoga April 4th 1865

Ever Dr Son

. . . A Sad acidant happened at the village yesterday. As they were fireing the cannon on a receiving a telegram of the Surrender of Richmond the cannon Burst and So injured John Stones Legg that he had to have it amputated above the nee. Dr. Squires done the Surgical work. This cast a gloom over the joy. Let all things be done with moderation. . . .

No pasture [pastor] at the B[aptist] Church yet. They continue to read H.W.B. Sermons. Col. Harris is Becoming more reconciled with those Sermons. Owing to the great fall in nearly all articles it Seems as though your expences ought to lesson Some. . . .

Your Father
Wm. H. Cook

LINCOLN'S ASSASSINATION

Ticonderoga April 19th 1865

Ever Dr Son

Called to Rejoice and mourn almost in the Same moment. This Should teach us how frail are all things earthly and draw us nearer to God in whom there is no change. The loss of our Presidant, O how Sad. I think this trajedy will tend to arrouse the people to more earnest work. H. W. Beecher Says Simpathy, and many other leeding men Say charity, and So do I but not to a Traitor. Put them down. Rub them out, until they will not go prowling about trying to kill Sick men. . . .

Friends all Safe in the army as far as herd from. All business Suspended today on account of the Presidants Burial. Rev. Bronson delivers a discourse tomorrow as a day of Thanksgiving and Prayer to God.

Gather up the knowledge. We need men in the field today, and Shall need them in the future. May providance protect us all.

Wm. H. Cook

THE BOYS COME HOME

Ticonderoga June 15th 1865

Ever Dr Son

As to a profession I have but little choice. Your tempre-ment is Souch that I have thought at times that Some thing besides theology would be best for you. You must choose that profession that you can go into with your whole Heart. . . .

The Soldiers of the last call returned yesterday about 20, or more. The Band and Col. Calkins met them at the landing with music and Speech and escorted them to the village. They are looking verry well but Sadly adicted to bad habits. . . .

Your father

Wm. H. Cook

CHAPTER V

Adirondack Murray

FROM HARVARD, Joseph Cook, as he now called himself, went to Andover Theological Seminary, where he became distinguished as a student of theology. It was reported that at examination time many crowded the classroom to hear him answer the professors. But he never became an ordained clergyman, although he was usually entitled the Rev. Joseph Cook.

After he graduated from Andover he offered himself as a supply preacher and filled numerous pulpits in New England. He also associated himself with H. G. Durant when that wealthy Boston attorney turned to evangelism before founding Wellesley College. Later on Joseph Cook acted as pastor of the First Congregational Church at Lynn for several months.

We now begin to find many references to the Rev. William Henry Harrison Murray.

In 1869 there appeared from his pen *Adventures in the Adirondack Wilderness.* It was the literary sensation of the day. Until that time little was known of the Adirondacks. Murray pictured them in all their beauty—lakes of pure water alive with fish, mountains and forests abounding in game. Such descriptions have now become the trite product of the advertising writer. In those days they were something new. Twenty years before J. T. Headley had published *The Adirondacks, or Life in the Woods,* but

Headley lacked the spark to set the world on fire. Murray popularized his subject. His book went into several editions.

Enthusiastic readers began to outfit themselves for the Adirondacks. An invasion started that swamped the few small hotels and stopping places then available at Saranac, Lake Placid, and Paul Smith's. This invasion was called the Murray Rush. It has not stopped yet. It gave to its author the name Adirondack Murray.

Joseph Cook and Murray were members of the class of 1862 at Yale, but no great intimacy appears to have subsisted between them until they both came to Boston about 1870. Murray was then pastor of the Park Street Church. He had been called there from a church at Meriden, Connecticut. Upon leaving Meriden, Murray recommended to the elders that they extend a call to Joseph Cook. O. H. Platt, who later achieved fame by means of the Platt Amendment, of which he was the author while a member of the United States Senate, extended the call. Joseph declined it.

Murray and Platt often went together on expeditions into the Adirondacks. Sometimes they camped on Lake George, not far from the Cook farm, or on the grounds of Fort Ticonderoga, at both of which places Joseph had visited them. Murray had visited the Cook farm, to which he was attracted by William H. Cook's stable of fine horses. The Park Street pastor had an eye for horse flesh. One of his literary works was a book called *The Perfect Horse.* We are now to learn something of Murray and his horse trades.

We shall also learn of Kitty Cook, the fastest mare William H. Cook ever raised, and the cause of no little excitement in the Champlain Valley.

MURRAY STARTS FOR SARANAC

Theological Seminary
Andover, Mass. Thursday, July 11, 1867

My Dear Parents,

. . . I enclose copies of two letters of mine to Mr. Murray by which you will see with what interest & pleasure I have watched the recent great increase of his means of usefulness by the putting into his hands, by Providence, of the weapon of the newspaper, the use of which by the pulpit I hope is yet to be made as effective as its use has ever been made by any form of the literature of mere amusement, or of error. Last Monday Mr. Murray Started for the Saranac Lakes to be absent some six weeks. In his last letter to me he says, "I am, beyond doubt, by God's aid, already a power for good in this section of the State, beyond what I had any reason to expect, indeed. That I work hard is true, & have for years. For the last few months I have the very delightful feeling that I am at length getting *command* of *myself*. My brain no longer plays antics, or goes by jerks and jumps, but is settling down to a steady pace. What I need now is time & health & God's sweet grace." I assure you that, remembering some conversations that Mr. Murray & myself have now & then had on the importance of severe mental discipline & on its methods, I did not read those words without emotion of a very pleasurable kind. He is aiding greatly the cause of Prohibition in Connecticut. . . .

Your affectionate Son,
Joseph Cook

IN THE PULPIT AT LAST

Ticonderoga Jan 20th 1868

Ever Dr Son,

. . . So you have entered the pulpet for the first time. I can feel Some of the Solemnity that must of attended this first effort. I Pray that you may be filled with the spirit of God and be able to teach the gospel truths Successfully. Hew to the line no matter where the chips fall. . . .

Rev Ashton and famely are viseting in town yet. He has been here eight days with his famely. He is Ashton yet. He no doubt is verry destitute. He would like to find a place to Preach. He has Preached twice Since he came here. His Sermons are instructive and delivered with energy. I do not Se why he cannot Succeed. I wish we could help him to a place.

There has been a Mr. Smith a Lawyer I think, from White-Hall, at ti to Start a Society of Good templers Something Simeler to the Sons of Temperance I Suppose. I have not attended their meetings. Clayton D[e Lano], Rev. Alling and Some others Seem to be verry active. They think that they will be able to get a Hundred or more to Sighn Ladies included.

They are talking Railroad Some and begin to enquire who can get ties. The Iron and Graffite is not working. Its rather lean times in ti just now. I have just paid my taxes, $222. This with what I have to Send to you keeps money matters pretty close. I hope you will be able to earn part of your expences now. . . .

I am reading the Bible more than usual, and Sermons. I am Sattisfide that after all the Preaching and teaching that the masses do not under Stand the great plan of Salvation thro the Atonement. . . .

We are approaching Spring. In my Leasure I am Oiling the harness and getting ready. Look well to your health. The Ship must be Sound to make a Successful voyace [!].

Your father
W. H. Cook

Tewsday Morn Jan 21, 68

Ever dear son, . . . Well it seems that you have fairly decided to take the Ministry as a profession. You have taken a great responsibility upon yourself, but as you think it is your duty,

I must try and submit my feeling to it, and pray that you may be the means of doing more good, than ever was yet done. . . .

I have lately been seeing some of the difficulties and trials of the ministry, which perhaps has prejudiced my mind against the profession more than usual. When I see a man with the talent of Mr. Ashton living around with a wife and 2 children from place to place wholey dependent upon the charity of others, I cannot bear the thoughts of a son of mine becoming a Minister. I cannot feel that it is Christianlike or honest, to live in that way. But they all seem very happy, and do not even ask whether it is convienent to have him and his family in your house from 1 to 6 or 7 weeks or not. Oh! my son shall I ever see you in such a situation? Pray God to strengthen me or I could not endure it. I have no objections to your being a Minister, but I *do want* you to be a different one than ever I was acquainted with yet. . . .

From your Mother
M. Cook

. . . Granma is very smart and well. I heard that they had a party of 25 couple at Mr. Barbers, a few nights ago, and Granma was so animated that she sit in her door to see them dance till 12 oclock. . . . You find enclosed a tooth pick, taken from a deers leg, a present from Robert.

PROSPECTING FOR THE RAILROAD

Ticonderoga, Feb 10th 1868
Tuesday Evening ½ past Eight

Ever Dr Son,

This after noon H. H. Moses brought the Advance and the Meriden paper with Rev. [W. H. H.] Murrys Sermon which I have just been reading loud to your Mother and Miss Thatcher. I like the Sermon becaus there is Something new or differant from the old roteen. Murry is bound to do good. I have been looking over a piece in the Advance on Free Masonry with a goodeal of interest as I wished to know the Religion of Free Masonry. I like the Advance. . . .

Johnathan Burnet died last Thursday evening. I attended his funeral last Sabbath. An Episcopalian from Middlebury officiated in a white robe. . . . He was burried in the new Burrying ground at Tuffertown. About fifteen Sleighs in attendance. . . .

The rout for the Railroad is not established yet. A few days ago they were looking along up trout brook as far as the J Cooks farm runs and thence northerly towards crown point. Some think it will go this rout, comeing thro the village near the Post office. . . . The Railroad, if it goes, will cost us Something. Let us proffit by Murrys Sermon—pay as we go. There is going to be no end to taxation.

I cannot be in my business with that phisical power that I have heretofore. I feel that I am becomeing an old man full of Aches and pain. The enjine is liable at any moment to Stop beeting, but I have an inheritance whither I am tending that fadeth not away. I can give you little advice. The cord that I have kept a constant hold of thro your course of Education Seems to grow less and less as you advance towards the close of your proffessional course. I feel that I have done my duty according to the best of my ability, and now I commend you to God. . . .

Your father
Wm. H. Cook

JONATHAN BURNET

Andover, Mass., Feb. 26, 1868

My Dear Parents,

The news of Honorable Jonathan Burnet's death moved me deeply. When I first began to go away from home to school, I always felt stimulated in looking back to him as the best educated man in my native town. Some of his conversations with me were of not a little service in quickening my interest in classical & historical studies, to which he was himself strongly attached. It was owing to information received from him that I was sent to Pointe-aux-Trembles to study French. I believe that it was he who pointed out to us Phillips Academy. I remember perfectly his giving me at his house a letter of recommendation to some one in Montreal on the occasion of my going to Canada. . . .

I never felt while he was living that I was, as a student, without sympathy in my native place, & it is an affecting circumstance to me that his grave is in our own valley.

Your Affectionate Son,

Joseph Cook

WITH MURRAY AGAIN

Andover, Mass., July 7, 1869

My Dear Father,

I spoke last Sabbath in Boston at the Bowdoin Square Baptist Church opposite the Revere House. . . . I happened to go to this church because I had applied at Mr. Sargent's agency in Cornhill for a place. The church stands on a square from which the horse cars for Cambridge start; &, while in college, I had often noticed its stone front. It was interesting to me that, the first time I ever spoke in Boston on the Sabbath, I should speak in a church facing Cambridge. . . .

In the afternoon I attended communion service at Park Street Church. My friend Murray I had never been able to hear in Boston before; but he spoke this afternoon only some fifteen minutes, & the audience, which numbered about a thousand, was smaller than usual, as no sermon was expected at the communion. After the meeting, my friend & myself had as much as two hours conversation at his study & while taking supper together at the Parker House. His book on the Adirondacks has just passed into a tenth edition. I wrote to him the other day asking him whether, while the book is giving him the ear of the people, he could not publish a volume on some of the most vital topics of practical religion. I asked him to publish, if his circumstances permitted, a book of which the object should be to put within those to whom his first volume had given sides that ache with laughter, hearts that ache for righteousness. I found that several others had made suggestions of the same kind to him; &, among them, Messrs. Fields & Osgood, who have offered to publish such a work for him. He is considering the matter; &, I think will publish, within a year or two, a volume of sermons. I found him very well tired out, as I had expected to do. He told me that he was living in a whirlwind; & that if he could have his own way he would drop into total obscurity for ten years, &, in that time, rest & ripen for his present work. He repeatedly said, with much emphasis, that he did not like his present life; & that nothing but a sense of responsibility held him in it. I came back to my fourth year at Andover all the more content with my present temporary seclusion & quiet, on account of our conversation. . . . My friend is studying scepticism as it appeared in Theodore Parker, & has recently given before his audience two extemporaneous discourses on his Impressions of Boston, philosophically & theologically considered. He has added $900 worth of books to his library since he came to Park Street Church. He was to leave Boston Monday morning for a two months' vacation in the Adirondacks. . . .

Your affectionate Son,

Joseph Cook

GRANDMA'S DEATH

Ticonderoga, Feb 27th 1870, Sun Eve

Dr Son:

. . . At four O clock, on the morning of the 24th of Feb. my mother ceased to breath. Her death was as easy as a childs

Sleep uppon its Mother's knee. I had Sat up with her alone all night. I knew that death was near. She had her Senses most of the time. I held her head and hand mutch of the night. She often Spoke my name with mutch Love. I Laid her back in her chair closed her mouth and eyes and held them for Some minutes, untill they would retain their place. Neighboring women were called in to lay her out. I Spent a moment with the famely as to the Arrangements for the funeral Services. I told them that I Should not dictate as to a Speaker. I told them that I wished Singing. George Said that he wanted as little Said and done as possible. I left then for home.

They called on Mr. Havens. He could not come on account of old age. So they called on Br. Mooney the Methodist minister. Rosina Selected the 15 chap of Ist Corinthians to be read without a Sermon and he made a Short prayer, Services opening and cloosing with Singing. There was but 15 mourners. 10 Sleighs went to the grave. The Day was beautifull Sunny, and Sleighing good. So one has passed away that Served as a Link to bind us all to geather. May the living proffit by this act of providance. I hope that Mother's faith was right. . . .

Wm. H. Cook

MURRAY BUYS HIS STOCK FARM

Lynn, Oct. 12, 1870

My Dear Parents,

. . . I met Murray ten or fifteen minutes by accident on Monday [in Boston]. We never meet without the Asiatic salute of arms under arms in the freedom of college days. I judge that Mrs. Murray's health is a source of some depression to him. She is now in the Adirondacks at Martin's, recruiting. He spoke to me, for the first time in his life, of intemperance as one of the public evils he resisted from an agonizing personal experience in regard to his own father, who does not now live with his mother, & is intemperate. My friend has bought the old homestead in Guilford, where he was born, & is putting stock on it. He seemed proud of owning nearly 200 acres of land. He is not to have a series of meetings in Boston this winter as he did last; but he is going to try the lecture field somewhat with the subject of the "Adirondacks." . . .

Last night in Boston I heard Thomas Hughes lecture on the relations of England to America in our civil war; & there is hardly another Englishman, except John Bright, whom Boston would have recd with more enthusiasm. . . .

Your affectionate Son,
Joseph Cook

BRIDGING THE LAKE

Ticonderoga, December 25, 1870

Ever Dr Son,

Christmas finds us all well. . . . Old Mr. Wetmore that lived with Mrs. McLaughlin was burried today. He was 94 yrs old had never manifest any faith in Christ. The mercy and goodness of God, O how great, Even towards his Enemies. . . .

I have commenced my rout across the valley to fodder. Elijah goes to School. Your mother does the work which has been quite hard for the past fortnight as we have been butchering. But we are now quite well tucked in for the winter. . . .

I wish that I could give you a more favourable account of a revival in our town. There has been but three Babtized as yet. Elijah Labounty was the first person that was Babtized in the Babtestry in the B[aptist] Church 2 weeks ago to night. Mary Benet and a Miss Nichols were the other two. Several have come up to high water mark and there linger. . . .

I hear that three men from Massachusets are in our town

to commence Bridging the Lake. They are boarding to the fort house. . . . I think we shall be Something in the future.

Wm. H. Cook

PLANS FOR GOING ABROAD

Ticonderoga Jan 16th 1871

Ever Dear Son:

. . . Bridgeing the Lake is progressing. They are to have $80,000 for building the Bridge, So I hear. . . .

I have looked thro Murrys Music Hall discourses again. After all plymouth pulpit and the Christian Union is my platform. . . .

I am glad that you have thoughts to your going Abroad. Great changes may take place within four months. I think the Sooner you go the better it will be for you. Its about time that you put a keen edge uppon all your tools. . . . Let the Lime Barrel of Slacking lime be your guide in delivering divine truth.

Your father
Wm. H. Cook

We have in the above letter a reference to the drawbridge that was later built across Lake Champlain at Fort Ticonderoga, a link between the Delaware and Hudson railroad and the Rutland, Vermont Railroad. This drawbridge remained in use until about 1930.

ACROSS THE RAILS FOR THE FIRST TIME

Ticonderoga Feb 19th 1871
Half past Seven Evening

My Dear Son

. . . I drove down the crick the other day on the ice round to where they are Bridging the Lak. I think they are Sinking the fifth pier. . . . There is a goodeal of opposition as to their bridging the Lake by Whitehall people but they keep on working. I drove off the Lake at the Deal place. They have laid the track across the road just far enough to turn and there Stood two pasenger cars and a rude rough board Depo. So I crossed the rail track for the first time in our town.

Your father
Wm. H. Cook

Years before this William H. Cook had planned that Joseph should study abroad. He sailed in 1872 and was absent for two years, studying in Germany and touring Europe and the Holy Land.

THE DESIRE TO GET EASY

Ticonderoga June 10th 1872

Ever Dear Son,

. . . I Shall Send a check of $600 to Bowles and Br., Boston, on the day this letter bears date. . . . I sometimes fear that we Shall loose money Sending it to you, you are So far from home. I advise you to be thorough in your German Studies. After masturing this Language I think you had better add more Speed to your travel. Try to get back to America in two years from the time you left. This will give you time to visit palestine and Jerusalem. You must get along as prudant as possible. I think the above Sum will carry you to Nov, I hope farther. Your letters are a great comfort to us as well as proffit and we are well paid for all in your good progress in your Studies. We Bless God for Souch a Son.

I have been oblieged to work verry hard the past Spring. I have not had Steady help. Elijah has left me. We had no words. He went home verry friendly and choose not to return. I expect another Boy Soon. I have no confidance in any one. There has been more leaks in my business the past year than usual. I Lost in Stalk and debts. . . . I must get out of this hard work if I have to Sell my farm. I am growing gray and Short in Step. I am not well adapted to manage hired help. Help has verry little principle. The great disire is to get Easy. This prevails in all our National affairs. I have accumilated a Small Sum by the Sweat of the brow. I do not mean that it Shall go to Support indolance. But this Strain of thought is not pleasant to you. I forbear. . . .

Your father
Wm. H. Cook

EVERYBODY WORKS ON THE RAILROAD

Ticonderoga Aug 18th 1872

My Dear Son

The brightest Spot in the journey of Life is your Success abroad. This Success in your life has ever been the elevating power in my life. . . . I Suppose that as Soon as the weather becomes a little cooler you will Travel on to the East over the Alps. Take time to gather up all that you want. . . . Write us a letter from Jacobs Well in Palestine that charming valley. I Saw Kimpton in our town the other day. He and his Lady were riding in a verry fine carriage with a Servant to drive. They Both Look older and care worn. He inquired for you. I told him that you was in Berlin Prusia Studying. He Said he once loved study but fortune had turned him another way. I told him that a mans life consisted not in the abundance of things he possesseth. He passed on. They say he is worth Millions.

I Suppose we may Say that they are to work on the R Road from White Hall to Ty. They have commenced to tunnel the fort grounds crossing at Roger Delanoes. Evry boddy is going to work on the R Road. Our town is growing a little but it will be a long while before we Shall get out of the ruts and run harmoniously for each others interest. There will be no rise in real estate at presant except in favoured localities. . . . What do you think of Greeley for presidant? Philips and Beacher against him, Sumner for him. There is a good deal of excitement in the coming election. . . .

Your father
William H. Cook

About H. H. Kimpton, Robert Selph Henry writes:

South Carolina fiscal affairs were further complicated through the employment of a financial agent in New York, H. H. Kimpton, a college classmate of Attorney General Chamberlain. Mr. Kimpton was surety on his own bond in the amount of fifty thousand dollars to safeguard the state in his handling of millions. He was given and used wide discretionary powers in disposing of the state's bonds through various channels at various prices, and in attempting to bolster up the state's failing credit by hypothecating bonds as collateral on other obligations of the state, on a basis of about forty cents on the dollar. Upon default, when Mr. Kimpton could put up no more collateral, the hypothecated bonds would be sold by the purchaser, with a further increase in the state debt . . . (*The Story of Reconstruction,* p. 371).

The state of South Carolina borrowed money through Mr. Kimpton in New York by putting up four dollars in bonds to one dollar received in currency. Interest was paid at the rate of eighteen per cent or more, besides commissions to Kimpton (see *idem,* pp. 444, 445).

THE FARMER'S LAMENT

Ticonderoga Aug. 30th 1872

Dear Son

. . . while I am writing on the back stoop, the rain is desending verry fast. I never expearanced a Summer like this, a great deal of thunder, hot weather, and rain. The flats have been flowed quite a number of times Since Haying. Hay and oats on the brook entirely lost. . . .

I cannot get help at any price. Men are flocking to the railroad to work for 2 & 3 dollars a day, and only eight hours labour. They begin to talk for eight hours labor for a days work on the farm. I have worked verry hard this Summer and accomplished verry little. My men are leaving me and going for higher wages and less work. I must loose largely because I cannot get help. . . . I cannot attend to my farm work as I used to. I must Sell part or let out but no one wants to take or buy. Land seems to be a drug. Farming is Slow business. Every one is Striving to live without work and most of them Succeed. My case is the case of all large farmers in our town. . . .

Your last letter presented a tone of Sadness or rather loneliness which gives us a little Sorrow on your behalf. You cannot rais men folded in tisue paper, the chiseling and grinding will give pain Sorrow and loneliness at times. Nevertheless it bringeth forth the perfect day. . . .

Your mother Says tell him to come home. I cannot feel that this is best. I think your next year, if you carry out your former plans, will be as usefull as any year of your life. Its going to cost quite a Sum, but what is Money to knowledge. . . .

The 2 of Sept. I am 60 years old. The 10th its a year Since you Shiped for Europe. The lord has blessed you and us. What will another year bring about the Lord onley knows. . . . My business gives me mutch unrest. I desire to have less to do. I want Some one to take the lead and let me have a little rest. I cannot hire it for money. I am the first to start and call in the Morning the last thro at night and this has been So the most of the time for forty years. . . .

I desire to help carry you thro and See you in business and then I Shall feal that I have in part finished my work.

Your father

Wm. H. Cook

STREET LAMPS IN TI

Ticonderoga Feb 2d 1873

My Dear Son:

Yours from Milan Italy Jan 6 reached us Jan 28th. I Suppose that you may be in Rome by this time. You are So far away that our thoughts are hardley able to follow you. . . . It Seems to us that you are not as Safe among the Italions as among the Germans. This is owing to our Simpathy for you more than for any real cause. . . .

Our famely is verry uniform and regular. We are all well. I do my chores and draw wood and drive colts. We are having a cold Snowey Winter. I am loosing a good many Sheep on account of the poor Hay fed to them. I Shall Soon have to look for help. I do not know where to get it. I will trust in the Lord.

Men are Surveying for a R. Road between the Lakes [Lake Champlain and Lake George] to run Some where near the foot of Mt. Defiance for pasengers.

I Suppose Burleigh Deposited $600. to your credit the first of Feb according to order. . . . I hope that this will carry you untill the first of June and that another deposit of $600 will bring you home. This will use up the most of our Surplus earnings for 20 years. We have the farm. We can live. We Shall never do mutch more than to live, now we are So old. . . .

There is more earnest work in all the churches this Winter. I am looking for a Revival this winter. This is my prayer. Business is quite brisk in the village. We now have Six lamps burning in the Streets on tall posts in the evening. The C[otton] factory Looks fine in the evening when lighted up. We hear the car and factory whistle quite often thro the valley. . . .

Your father
Wm. H. Cook

FARMER COOK'S HARVEST BEGINS

Ticonderoga Dec 14th 1873

Ever Dear Son

Your two Letters have arrived Safely and by their contents we learn that you are Settled in Boston and have Set the ball roling to your own Satisfaction. . . . You have gon to Boston with a good Set of tools costing some $15000, to put them in good order. Besides mutch care and anxiety for your success, we have been greatly blessed. Our efforts has not been a failour. I shall expect to hear of some good work from you. . . . You have health of Body and of mind that few Students possess. You are also free from all bad habits. You ought to be a power in the world to do good. You have no reason to doubt your ability to do what ever you choose that is in harmony with the will of God. . . .

I am mooving round among the flocks doing chores and keeping things right. The Horses are doing well. Nothing new in the Kitty Cook Suit. The trial may not come off untill May. If they drive her fast I Shall put an Injunction uppon her or take Some other course to keep her quiet. . . .

As regards your Lectures uppon places that you have viseted abroad Let them be full of detail. Not one in a hundred knows anything about those places. Dont doubt your ability as a lectuerer. Let the Black board be freely used in every manuit point. The Black Board is better than a Map for your use. The audiance do not become confused as uppon a Map and you can lead them along as you choose. I hope you will Soon be able to Spend three evenings if not four out of a Week to your own Satisfaction and to the benefit of the audiance that you address. I have Set my face to get out of debt as Soon as possible and to keep out and I advise you to do the Same.

The Hammer and Saw are Still herd in our village. . . . The Spindles hum every day and business moves on. The Railroad men are prepairing to build an extensive dock just at the turn of the road as you go from port martial to Putnam costing one Hundred thousand dollars and covering two or three Acres of the creek or Lake. This will be the head of navigation. From this dock goes the railroad to the Sounding waters. . . .

I am watching Mr. Beechers conduct closely. It look a little as though he would go down in the Scale of my estimation. . . .

Good bye, the Lord bless you and Enlarge your field of labour. The world is the field go work therein.

Your father
Wm. H. Cook

AGASSIZ AND MURRAY

Boston, Dec. 18, 1873

My Dear Parents,

One thought gladly of Agassiz's silent prayer before his audience at the dedication of the Penikese Island School a few months since, when today through the hushed crowded Appleton Chapel at Cambridge he was borne in and out in his coffin. There was no address, only a most impressive selection of Scriptures, a prayer and solemn hymns. Sir Edwin Landseer, whom I saw lowered through the floor of St. Paul's Cathedral to his last rest, was not buried with less ostentation. But a great number of men of science were present; the chapel was draped on the interior with heavy crape, evergreen, myrtle and the appropriate flowers; I saw strong men with suffused eyes carrying away as sacred mementoes pieces of the myrtle sprays. Boston, I hope, will not suffer the Penikese School for teachers of Science to fail. The great Cambridge Museum is already equal to any of its kind in Europe, and will be the chief instrumentality by which Agassiz's life will affect the future. But there is no man left in the country who can draw money for learned pursuits from the mercantile classes as Professor Agassiz did. He had a slight paralytic shock years ago, but did not heed the warning carefully enough, and a second shock proved fatal. At Lake Neuchatel in Switzerland 66 years ago, this great career began; today it closes in America among the groves of Mt. Auburn. From my place at the summit of Boston, I look east beyond the harbor lights on the shoreless ocean; and west beyond Agassiz's tomb on shoreless eternity.

. . . I have seen Mr. Murray, took dinner with him at the Parker House by his invitation, & afterwards conversed with him three hours in his study. My impressions are that his heart is sound; & that, in spite of a constant whizzing of criticisms & attacks from cultured sources in Boston, his general hold upon the middle class of culture here has strengthened in the last two years. The publication of his sermons weekly was continued only twelve months. The fact that Boston hears him preach makes his book on the horse seem other than the best half of him; I think that the book, however, has perhaps thus far injured his influence. A new edition of it is to appear soon & I will send a copy to my father. Mr. M. has had a slight paralytic shock, & talks of taking a year's rest. He remembers my father with much affection; & I found he knew more of my fathers horses, including the one now in law, than I do.

Your affectionate Son,

Joseph Cook

Farmer Cook's horse "now in law" was Kitty Cook, the mare referred to in the letter preceding this one. Later on we shall be privileged to follow the history of this animal and the litigation over her.

MURRAY BUYS A COLT

Ticonderoga March 2d 1874

My Dear Son,

. . . Wall, you have Introduced yourself in the Shamplain valey. Mr. Harris Says that they Say at Middlebury that your Lecture was too deep for the most of the audience. . . .

On Monday, the 26th Feb, Rev. Mr. Murry and othher gentleman from Boston were here with Mr. Harris, to look over the Horses. They Seemed to be well pleased with the Horses. I Sold Murry the colt that was in the team that carried you to the Lake for $400. He took him away with him. Is to Send check on reaching Boston. Murries wifes Sister was with him here. Murry says that he is going to get you to run his church for Six months So that he can get a little rest. He Seems full of life and Horse. Look well to your own interest for the future. Be careful what baits you bight at.

Since Murry left, Mr. Harris came over to See me and brought a letter from Murray Saying that He wanted Old White at $250, and the cosset colt for $800. They all went to See Kitty Cook. Speak verry highly of her. I told them that I desired to get out of Horse breeding on account of the excitement. They Said that I must not think of any Souch thing. Said that I was the best breeder of Horses in those parts. Murry Says that he can help me to Sell all the horses that I can raise. How far it will do to trust him I do not know. Murray will be here again Soon with other gentleman. I hope and prey that Rev. Murray will never fall from his high Standing as a Spiritual teacher. The Lord bless him Abundantly.

The tinkling of the bell as regards the Kitty Cook Litigation is drawing men from all parts to See her. Mr. Harris Sais that he brought over three men from Boston to see Kitty. One of them asked to drive kitty a little and was pirmitted. After looking her over he Says that She is the best mare that he ever Saw. The Horse click, my oponants, are Sure to hold the mare. Time will tell. . . .

Your father
Wm. H. Cook

Dear Son . . . Your Pa is better, and is full of Horse trade. I suppose he will sell old grey, and I feel as if I could not have it so. I would like to give you a discription of Murry's call here but have not room. His wife's sister is as much horse as he is. She went to the barn and over the brook with 5 men to see horses, and could talk horse as well as any of them.

Your Ever loving Mother

M. Cook

MURRAY PROPOSES PARTNERSHIP

Boston, 9 Ashburton Place
March 5, 1874

Dear Parents,

. . . Deacon Farnsworth saw me on Saturday at his house & I saw Mr. Murray yesterday by Mr. Farnsworth's request in regard to the plan of my occupying the Park St. Church pulpit in April & May & perhaps in June. . . .

Murray told me he had a great desire to form a kind of partnership with my father in horsebreeding. He thinks he could take my fathers young horses, train them in Connecticut on his farm, & sell them in Boston or New York greatly to my fathers advantage & his own. He spoke savagely against the Ticonderoga & Whitehall horse rings, the evil reputation of which he said was well enough known in Boston. My father, he maintained, had raised more excellent horses than any other man in America. The Cook farm he thought likely to be remembered in horse literature longer than he or I should be for anything we might do. He had made an offer, he said, of $15,000 for Kitty Cook, to any parties empowered to give him a quit claim deed of her. He told me what three horses he had bought of my father, & seemed pleased with his bargains. I listened to all this without much comment; but, if my father needs information about Mr. Murrays farm in Connecticut or other matters connected with propositions for uniting the business of the Cook Farm with that of the Murray Farm, I can, I think, get at the essential facts without danger of important error. . . .

Your affectionate Son,

Joseph Cook

HEAVEN WITHOUT HORSES

Boston, 9 Ashburton Place, March 26, 1874

My Dear Parents,

. . . After the newspaper gale produced by the criticisms made on Murray in his church, I saw him & Deacon Farnsworth separately. Murray was evidently annoyed, but in good spirits; and the Deacon said that he had received any number of letters endorsing his remarks as printed in the newspapers. There are two parties in the church & Murray thinks his side is much the stronger. I hope the difficulties will be settled; but I should not be surprised if Park St. had another pastor in the course of two or three years. Probably Murray has been careless; he has visited little; his farm and his lectures take time his people ought to have; but his sermons everyone seems to consider excellent. I urge Murray to go forward & have a revival as the best reply to all criticisms, & not to plan for leaving either the ministry or Park Street church.

He knows very well that I think he makes too much of horses. Heaven to me will be perfect, even if it does not contain a racecourse. . . .

Your affectionate Son,
Joseph Cook

CAN'T WE TRUST THE CLERGY?

Ticonderoga May 4th 1874

My Dear Son

Yours of the 30 April was Recd. It gives me joy to hear of your Success but if all that you do only pays for your Board, whats the use? People of less tallant do this. I feel a little like Scolding but I forbear.

I have been disapointed in many ways this Spring. Rev. Mr. Murray pays no attention to the traid he made with me for horses. He promice to Send check as Soon as he reached Boston. That was two months ago. No check comes yet. Can we not trust the clergy? No, it Seems not, but I wont Scold. I will try to help you to the money to pay your Andover debt. The kitty cook Suit may come off 1st June. If I get beet it will cost me a large Sum. If it goes in my favor it will Still cost money. Your call is unexpected. I am use to Souch things. I will do the best that I can. . . .

The Sun has Scarcely Shown for the past week with a terable wind continually. We have two colts, lost one. Shall have two more Soon.

Your Mother is not quite so well. Your parents have got about thro. You will have to Stand alone Soon; I do hope that you will not enter into any more business that will not part pay expences. Is it a Sin for one to Strive to pay his weigh? I think not. . . .

Wm. H. Cook

NOTHING RIGHT BUT JOSEPH

[Undated]

Dear Son

Your father Says there is nothing goes right but the boy. He is right every time round. When we expect anything from him it always comes. In this he means your letters, I suppose, for they come So regular and are of so much consolation to him. But your last letter makes him rather Sad for he thinks you are in want, which is not a pleasant thought for him.

He has been So disappointed in Murray['s] proceedings about horses so that he begins to think there is no confidence in anyone. He has not Sent pay for the horse that he took, neither has written him but once, and then said he Should be here Soon, but has not. *Old Grey,* one that he bought, has lost her Colt, and he is afraid he will be blamed. I hope he will not take her. . . .

Why not come home and Stay a while with us, if you are not doing any thing more than paying your board? We will do that for nothing, and you might not have to do So much for nothing. I wish you was not in debt for that I think is one of the worst things, for a man to not be able to pay his debts when he agrees to. You know that I never believed in any-one's doing any faster than he could pay. . . .

I wish you had a good home of your own, and some one to take care of you and help you, and not pay board nor room rent. . . .

Your affectionate mother,
M. Cook

THE KITTY COOK TRIAL

Ticonderoga June 7th 1874

My Dear Son

All well. Last Monday, June 1st found myself and witnesses at Elizabethtown. The Kitty [Cook] case was taken up tuesday afternoon. I was on the Stand as a witness Six hours, under the examination of Judge Hand, Baile[y']s lawyer. Wednesday finished the examination of witnesses, and they went home in the evening. Great excitement prevailed. Thursday the pleas were made by Hand and Grover, a pardner of Waldo and Tob[e]y. At 11 oclock the jury retired to consult among themselves. At One returned to the court room all agreeing in a verdict in favour of plaintiff for the return of the Mare Kitty into my hands, or $8000, if I Should not be able to get the Mare Kitty. I returned home in the evening by Stage and boat. Found Elijah at fort to bring me home. Arrived at home just daylight, verry weary and tired.

Found Rev. Mr. Murray, Dr. Quint with him, and a gentleman from Middleburry. No rest for friday. I looked over the horses in fournoon and they left in afternoon for Middleburry. Murray took the cosset at $800. Gave his Note payable on demand. He Sent check for the horse that he took away when here before, $400. Did not take the Old White Mare. He Says that he wants me to Send my horses to him as fast as they are ready for market or before even at any age.

Saturday. I hear that the Dr. is going to apeal the Horse Suit. I received Notice from my Lawyers to be at their rooms Monday the 8th without fail. No rest for me, but I must Say that a cloud by day and a pillar of fire by night has led mee thus far. The Lord is with me. . . .

I am at great expence and verry Short of money. I Shall Soon reach the wool which I hope will carry me thro. . . . The farm and Stalk prospects Look fine.

Your father,
Wm. H. Cook

PRIVATE

One word about Murray and Dr. Quint. I studied them close and the most remarcable thing that I found out was that they Seemed determined to Smoke all the tobacco that they could Smoke out of cob pipes which was constantly in their hand or mouth. If God be God, Serve him. If Baal, Serve him. No emalgrimation of the two. I fear that Murray is growing quite as fast in the Horse Spirit as in the Spirit of Christ.

I questioned them closely about you. It Seemed to me that they answered every question with a Spirit of rivalry. The Dr. Said that you had Brains anough but that you had better be at home on the farm. Murray Said that it was a great pitty that you was not located permanantly Some where. Said that you was able to draw a great Load but Lost mutch influence on account of your extensive field. Plow the Smokers under without looking behind. . . .

I will Soon write you what course the Dr. [Bailey] takes.

Great excitement. Many take off their hats and bow as I pass thro the Streets in town in a waggon. I do not feel to retaliate. I Shall keep a close mouth. No man was ever hung for not Saying anything. . . .

Wm. H. Cook

Let the eagle Scream. I had 5 Lawyers. I emploid Robert Hale of Elizabethtown, the ablest Lawyer in the county.

THE KITTY COOK CASE—The action of William H. Cook, Plaintiff, vs. James H. Bailey, Stephen L. Wheeler and Thomas E. Bailey, Defendants, was started by the service of a summons and complaint, the latter verified December 27, 1873. The action was in replevin and the complaint demanded possession of the "four year old bay mare with a white strip in her face known by the name of Kitty Cook," or damages in the amount of fifteen thousand dollars. It was alleged that "said mare had attained great speed" and that there was danger that the defendants, who were alleged to have possession of the mare, would drive her on the ice to exhibit her speed and thus effect a sale of the mare for a large sum. It was further alleged that "such use endangers her valuable qualities" and that the animal might be injured while so driven. Waldo and Tobey of Port Henry, a law firm of great local renown, appeared for the plaintiff. The equally celebrated Richard L. Hand of Elizabethtown put in an answer for the defendants. The answer alleged that Cook, on June 30, 1873, had sold the mare to the defendant James H. Bailey for $700 on the condition that Bailey might try the horse for a reasonable time before making payment. July 12, 1873, the defendant alleged, he notified plaintiff that he would keep the horse, and proferred payment, which plaintiff refused. The action came on for trial at Elizabethtown, June 2, 1874, before Hon. Judson S. Landon, justice of the Supreme Court, and a jury. The trial lasted until June 4. A non-suit was granted as to the defendants Wheeler and Thomas E. Bailey. Augustus C. Hand summed up for the defense, and Marcus D. Grover, a member of the firm of Waldo and Tobey, summed up for the plaintiff. The jury returned a verdict in favor of the plaintiff for the return of the mare or the payment of $8,000 damages.

Marcus D. Grover, whose persuasive eloquence was thus successfully exerted in behalf of William H. Cook, later became attorney for James J. Hill, "The Empire Builder," and for many years defended the negligence actions brought against the Great Northern Railroad. Grover was succeeded in the firm of Waldo and Tobey by Chester B. McLaughlin, who later became a judge of the Court of Appeals of the State of New York (1917–1926).

CHAPTER VI

On the Downward Slope

OLD AGE and a life of toil began to tell upon William H. Cook. But amidst the trials and disappointments of this life he found consolation in the success of his son.

PLOT TO RUIN BEECHER

Ticonderoga August 16th 1874

My Dear Son

Yours of the 10th and the paper came safe to hand. I was glad to hear you So well Spoken of by the Press. Your Mother and I and Kit have been to church today in the new carriage. . . .

I Set up last night until 11 O clock reading Beechers Statement before the committy. The fountain is being clensed. I drink more freely. His feelings are verry Simelar to mine at times. What a plot to ruin Beecher. Falshood, Blackmailing an[d] other work of the evil Spirit. I am astonished that it Should give him So mutch unrest. I have more charity for myself when I learn that So great and good a man Should be led through Souch a labyrnth of trouble and Still be able to perform So mutch labour. Yale Lectures, Life of Christ, and church besides all his misalaneous duties. The Lord bless him and make him doubly usefull is my Prayer.

No theaf has Stolen clif Seat nor double Vista. This is property that the world cares little about and needs verry little watching. . . .

On my return from carrying you to the Lake I bought a volum Sermons by Rev. T. Dewitt Talmage which I am turning over in my Leisure moments with pleasure. His discourses are to the point and full of the Spirit of God, well fitted for to elevate the masses. . . .

I go wearily thro Life. I believe that I have the Spirit of my Saveiour with me constantly. Pray for us as a family that we may have peice that comes from a pure Life.

Your father
Wm. H. Cook

ADVANCING YEARS

Ticonderoga Sept. 2d 1874

Ever Dear Son,

To day I am Sixty two years Old. I have pitched and mowed 6 large loads of Hay besides doing other work. The hay is Second growth. I have no one but Elijah with me now. He has been to Middlebury a week on a visit. I got along alone while he was gon. Henry & Roger delano are just finishing haying for the first crop. I have turned the cows uppon the meadow. . . .

I am afraid that H. W. B. is in the mire. Clergaman, be ware.

I am So tired to night I will finish this Letter in the morning.

All well this morning. . . . I am almost tired and discouraged with the battle of life. I want a change, Some one to take the labouring oar, but I must work on. The Lord help me. I think you have done exceedingly well in your five Sabbaths in Boston for which I am verry thankfull. I hope that you will be able to make your business pay as I am running behind at home. . . .

Your father
Wm. H. Cook

SEE-SAW WORK WON'T DO

Ticonderoga Sept 16th 1874

My Dear Son

. . . Was glad to hear that you were called to Speak at Cambridge on the old familiar ground to an inteligant audiance. . . .

Our famely consists of four now. Things moove on just as they did when you were at home. We do not vary 5 minuets in our rotene of exercises. While I am writing the famely are reading by Lamp light arround the table. I thank god that I have got a home, and that I have been permitted thro him to raise and Educate a Son that bids fair to make a man. Do your best and dont doubt your abilety to do great things for the caus of truth. Put on the whole armor.

The more that I am brought face to face with men the more I doubt evryboddy and myself to. I would have no confidents if I were you. Think of Motton. I think that I have been about as Soft in the management of Kitty the Horse as H. W. B. has been with Tilton and M[otton]. I can raise more Horses if it be gods will easier than I can fight for my rights in K[itty]. . . .

I Pray that a verry Large field may be opened for your comeing labours the comeing Winter. . . . I would put on about two more Horse powers to all of my exercises. Nothing but an earth quake or a thunderbolt will arrouse men from their Stupor. This Se Saw work wont do it.

The cunning Smart man as the world call them are leeding the masses of today. The most of the human family are after gain in wordly things. . . .

The R R to Lake george is nearly compleated. . . .

I think that it's thro Butler T[ilton's] councel that is giveing the B[eecher] case an unfavourable look just now. God will bring truth out of darkness. . . .

Your father
Wm. H. Cook

JOSEPH WRITES POETRY

October 1St 1874

Dr. Son

I think that you will have plenty of business now the weather is cooler and we begin to have evenings. Get your tools Sharp So that you may work with eas. All well at home. No change in the program. If you See Murray ask him about the yerling colt. Ask him if he wants the Bennet Mare that he drove when here. Tell him he had better buy kitty cook. The cars or gravel train is running between the Lakes. The town news you get in the centinal. The forest just begins to change. Not many leaves fall yet. Do your best. The Lord help you.

W. H. Cook

Thanks for your verses on lonelyness. They do me good. I can hardly understand your poetry in the Atlantic.

THE FARM IN AUTUMN

Ticonderoga Oct. 22d 1874

Dear Son,

. . . I See that you Boston men are after Beacher. Can it be possable that he is guilty, and Still carry forward his church with So high a head? I have not renewed for the Christian Union and Sermons. I am going to weight untill after the decision of the people. . . .

The potatoes are dug. Last week we thrashed. The corn is in the barn ready to husk in rainey weather. We Shall commence to plow tomorrow. We drive colts Some. The forest leaves are most beautiful now. . . .

I See by the papers that Murry is to Lecture at Platsburg

on Poverty. I hope that you will be called west this winter to Lecture. . . .

Wm. H. Cook

LIFE'S TAPER BURNING LOW

Ticonderoga Nov. 8th 1874

Ever Dear Son:

It is Something to go to Boston and be able to pay one's way in the business that you follow. You will find more to compete with as you go more into public life. Some to praise, Some to criticize. Life is a warefair. The man that wins must work.

I sometimes think that your efforts in writing and Speaking is to deep for the people. They want Something that has got a handle that they can take hold of. Give this a thought. . . .

I hear nothing of late about Murray. I believe that the congregational church intend to get him to Preach the dedication Sermon on the opening of their new church. They tried to get him to come here to lecture but could not, his time being wholey ocupied. . . .

With me this has been a year of trial and Sadness. I anticipated mutch peace after your return from abroad, but there has been more Leacage in business than ever before. I have had more difficulty to make the ends meet than usual. I have had more unrest. I have tried to do right and have continualy done rong. . . . I am constantly Set uppon by beggers for Some cause. I hope you will be able to lay by a little this winter for a time of need. I would not enter into any enterprize that would envolve you in debt. I can do no more for you at present. The tables may turn in a year or two. I may be able to Sell Something that will make me more easy.

What's the use of whineing when we all have health? Let us go forward trusting in god.

All well. Five in the famely. All Set around the table reeding as I write Still as mice. . . . How is it about Beecher? You Boston fellows are Still just now. Beacher is out Lectureing.

Mass. gon democratic that state of Brains. . . .

I grow clumsy every day. I Shall Soon be thro. I have done my duty the best that I could. Gods will be done. Your Letters and Success in life are the Sun Shine in my pathway. . . . It is now eight O clock. We will have famely worship and retire. Good by.

Your father
William H. Cook

AN AIR CASTLE CRUMBLES

Ticonderoga Jan 21St 1875

Ever Dear Son,

C. W. Bailey of M[ont]Peliar, Vt. and the Dr. James H. Bailey called on me by my request and we had a talk about Kitty which resulted in my Selling my interest in Kitty to C. W. Bailey for $1,000, he paying all expences on her Since June 18th 1874, which has been about $2,000. So crumbles the air castle of $15,000 down to one thousand Dollars.

I think uppon the whole that this was the best that I could do under the presant circumstances. I Saw that there must be a great deal of Litigation to enforce the calkins contract on account of its Looseness. I can rais another horse cheeper than to fight for my interest in her. Calkins has Shown his kindness to us by his intrest in the Dr. that has taken $10,000 out of us. Pleas remember this to him in the future. . . .

My next Son I am going to educate for the Law.

Your father
Wm. H. Cook

(Kitty has been injured in the races.)

(I hear nothing from Murray. I expect to have trouble to get my pay of him without Law. He wrote me that he thought he Should be at my place in Febuary.)

WEARYING OF HORSE TRADES

Ticonderoga March 8th 1875

My Dear Son,

I See by your letters that your field for labour is enlarging and that you are haveing more calls for Sunday preaching. . . . I See that you are drifting towards Andover and perhaps towards a proffessorship at A. But this would not quite do. I think you had better Spend 5 years in Subsoiling New England for the Gospel Seed. . . .

We have done verry little but chores and drive Horses the past Winter. I am Sick of the Horse business. There is to mutch competition and excitement about the business. It leeds one into the Society of dishonest and trickey men. The Horses have done verry well the past winter. I cannot Sell any Horses nor get pay for what I have Sold. Where is W. H. H. Murray? He Said that he would be here in Feb. I have herd nothing from him. Has he got his big free Church built in Boston or is he up to his farm yet? . . .

Your mother is cleaning House a little. She expects Rebecca G. and famely tomorrow. She is passing to and fro from Kitchen to parlor brisk as a girl. Mrs. G. thinks of Staying with us Some days. . . .

Here comes Charles L. He is right from town. He Says that A. P. Wilkie has failed for $35,000. Wilkie and famely have been Sick for Some time. I am on Wilkies paper for $400 Bank note. He Said that he would take care of the note. I presume that I Shall have it to pay. The large fish are already after the Spoil. I Shall See him in the morning.

I have just called on Mr. Wilkie. They are all Sick. I am in the boat for $400.

We are after the rummies with a Temperance bord of excise by 80 majority. . . .

Laura is ironing. Ma and Kit are dressing chickens for the expected company. . . .

What say you knowing ones in Boston about H. W. B.?

They have to have a poliece to keep people from rushing into the church to Suffocation, even in a Prayer meeting. I do not think that he will be convicted by the Jury. . . .

Your father
Wm. H. Cook

TI VILLAGE BURNS

Ticonderoga March 31st 1875

My Dear Son,

About 3 Oclock this Morning the Heart of our village was burned about 28 Houses.

I went down quite early to get a Letter from you. The Sight was Sickning. The Streets were full of people, everything in confusion. The Smoke was just passing away.

The four corners is entirely burned up from Hiram Wilsons house to Pincheons B[lacksmith] Shop al gon. The Brick Store and all up to C. Shattuck's, the whole of the tavern corner and Mr. Clarks House, the Maconic corner from Ramsays to Clarks Hardware Store all burned.

This loss and the Wilkie failure is verry bad for the town. . . .

Your father
W. H. Cook

MURRAY STILL IN ARREARS

Ticonderoga April 25th 1875

My Dear Son

. . . I wrote Rev. W. H. H. Murry a few days ago asking him to pay for the colt. He may Say that he never bought a colt of me. You need not Smut your fingers by having anything to do with the matter. I will attend to Murray. Look thro his napsack closely. Sound him well So as to report at Some future time.

And now I Say to you, down with the Subsoil Gospel Plow. Turn up the virgin Soil for the gospel Seed. I tell you the most of our Religious teaching is but pitta pat work. . . .

Your father

Wm. H. Cook

A great Success your Centenial at Concord. Al papers are filled with your doings. . . .

This last refers to an address Joseph Cook delivered at the Centennial Celebration of the Battle of Lexington.

MURRAY MAKES GOOD

Ticonderoga Oct 25th 1875

My Dear Son

. . . I hope that you are not Selling your labour to cheep or giveing away to mutch of your time. You ought to earn a hundred dollars a Week to make the ends meet in the future when you have more famely. Take precious care of your helth as you always have judgeing by your wt.

Our fall harvesting is closed up. The cellar is Stored with Apples and roots. Murray Has Sent a check for $800 payable in four months from 18 Oct 75 on Maverick Bank Boston. Says he will pay the Intrest when we meet. He has acted manley. I Shall hold the note until due for fear that it may be protested and make me trouble if I Should get it discounted.

We thought by your changeing rooms and going to F[air] Haven that you was going for a Housekeeper. Perhaps it may be So. The Lord bless you.

W. H. Cook

It was becoming apparent that Joseph was soon to wed Georgia Hemingway, after seventeen years of courtship.

READY FOR THANKSGIVING

Ticonderoga Nov 24th 1875

My Dear Son:

. . . I See by the paper you Sent that the enemy know that you are in the field. Drive them to the wall. . . . Moody & Sankey have done a great work at the Rink. How Simple and earnest they work, all their talk Biblical. . . . I hope that you may do a great deal of good in your labours this winter. . . .

I see your name frequently in the papers. I feel that you are making a Success in your labours. Be plain and Gentlemanly. Don't fear to tell the whole truth. Expose men in their hants of vice. Bring the truth to bear uppon them. I dont think that you will be troubled for work in the future. Boston is a great field for a young man to Succeed as a Speeker upon his own merrits. This it Seems that you have done. I do rejoice and thank God for your Success. I hope that you will not Sell your labour to cheep. . . .

We are all well and ought to be happy. We have an abundance of every thing but money. That is hard to be got. . . . Have not Sold my wool yet. No Sale for Horses nor any thing that I have. Robt and I do chores, Halter break colts and get them in Shape for winter. . . . I have 19 Horses & colts to winter, 15 head of cattle & 100 Sheep to care for this winter. . . .

Whats become of Murray? I hear nothing from him; tomorrow we Shall think mutch about you, it being thanks giving. The chickens have been dressed this evening by Laura and every thing is ready. We ought to be very thankfull as a famely for the past mercies of God. . . .

Your father
Wm. H. Cook

CHAPTER VII

The Monday Lecturer

THE Boston Monday Lectures began in 1875. They were at first introduced under the direction of the Young Men's Christian Association for the benefit of clergymen in and near Boston. It was not long before the Meionoan, a basement vestry of Tremont Temple, where the lectures were first held, became too small for the audiences, and when Tremont Temple was used Joseph Cook rapidly succeeded in filling it. Back in Ticonderoga, William H. Cook now felt that all his patience and toil had been rewarded.

JOSEPH AT THIRTY-EIGHT

Ticonderoga Jan. 26th 1876

My Dear Son

This is your birth day, thirty eight. . . . I am thankful that my life has been Spared to See the fruits of my desire in your Education. Two years more judgeing from the past will enlarge your field of Labours So that it will take the New England States and perhaps N york and the west. . . . A hundred Ministers evry Monday to talk to how rapped your acquaintance. Do your whole duty . . . , Studdy Boston with a keen eye that you may learn human nature and be able to apply the knife to the Cancer every where. . . .

Today I suppose you are at F[air] Haven. May your joy fill your anticipations and may God clasp you both in his arms and make clear to both of you the road that will lead you to mutual happiness. . . .

I like exceedingly the Remarks of the Boston Cong[regationalist] as regards your Noon meetings in Boston. . . .

Try and have a little Money at intrest to use in a day when you are not So prospered.

Your father
W. H. Cook

CAN'T PITCH OVER THE BIG BEAM

Ticonderoga Dec 11th 1876

Ever Dear Son,

I hope that you will be able to keep the noon Lectures up to the presant Standerd. Almost all our papers have Something to Say about your Noon lectures. . . .

Perhaps I Shall come to Boston when Moody and Sankey get well warmed into their work. Don't look for me untill I come. I thank the Lord for your Success and Pray that it may continue. . . . I am expecting that you will be called to N. York Soon. . . .

I cannot go up the Ladder as I used to do. . . . I have done pitching Hay over the big beam. I have Seen the time that I could get round with any of them. I am done. You will never know how I rejoice in your Success. Thank the Lord for leedind [!] me to do what I could to give you an Education.

Your father
Wm. H. Cook

SUNDAY IN WINTER

Ticonderoga Jan 15 1877

Ever Dear Son,

. . . Your father Kit and Moses have gone to meeting, and I do not know as they will get back home today, for the wind is blowing a perfect gale, and the snow flies so that you can not see the barn. . . .

We worry a good deal about you. We hear of so many disasters on the railroad. But your father says the Lord will take Care of him. . . .

Your letters and papers help to pass of many a lonely hour. You would laugh heartily if you could hear your father and see how animated he is sometimes after reading your lectures and some things that he sees in other papers. I often have to tell him that he acts silly, that his anticipations are to great, he should think how soon they might be blasted. You are his life, and always have been and have been a good boy, and may you be spared to ease him down the rugged path of old age which is fast upon him.

I am glad you are doing so well this winter. We often say to ourselves (for we are alone most of the time) that you must be earning money enough to support a wife if you saw fit to have one. How is that— . . .

There, they have to come, and I must stop and get supper. . . .

Mother

I never expected to live to hear of your filling a Station of usefulness So prominant among a cultivated audiance in Boston as you now moove in. I feel to give god the glory. You are truly bringing out the imitation of the Lime Barrel that we talked of under the old Shed years ago. . . . May you Become the Paul of the world the verry chosen vessel indeed.

I like the ring of your last Letter. No more Lectures Short of fifty Dollars. It ought to of been Seventy five dollars, or one hundred occasionally. I dont want to urge you to become a misor or Seek after money. The time is comeing that you will want means to do good. I do not forget for a moment the worth of Souls. . . .

It's a favourable time now to buy government Securities if you have means. Only about 8 cents premium. . . .

W. H. Cook

BREAD CAST UPON THE WATERS RETURNS UNTO HIM

Ticonderoga Feb 25th 1877

My Dear Son,

Last Sunday as we came from church we called to the office and got your letter with the check in it. Today as we came home we got your Telegram of the 24th. Was exceeding thankfull to hear of your continued good health.

I put you down as about four Men Strong. After you pass this point of Strength look out for pebles in your Shoes, or buckles in the harness that gall.

The papers keep us well posted as to your work. I was pleased with a piece in the Congregationalist Saying that you gave your Lectures extempor. You have now doubled the cape of the Moody meetings and are now in fair Sailing. I See no breakers ahead now but overwork. . . . I have read your Lectures with great intrest and proffit. Most of them I read twice. I have made many Sugestions about them that I have Seen afterwards in the papers. I think that you are doing a great work. . . .

I am sorry you Sent So mutch money. I did not think of dunning you. I dont want my record of furnishing in the past mared in the least. You Shall have the money on demand with Interest although its a great convienance to me just now. . . .

W. H. C.

TEN THOUSAND BY NEXT YEAR

Ticonderoga March 19th 1877

My Dear Son,

We had quite a laugh over the comic pictures in the Boston times. Its not evry man that can obtain So mutch room in So important a paper the first Shot. When men shall revile you for my Sake rejoice and be glad for great is your reward in Heaven.

I am exceedingly pleased with your Lectures. Allowing me to be the judge there is an improvement as they increase in number. I read most of them twice without tireing, and that is more than I can do with Beacher Spurgeon or Murray. . . .

If half the Press Says about you is tru you ought to have Ten thousand for your next year, Say Eight for the Noon lectures, and Two for outside labours. Anything less than this I Shall not be Satisfide with. . . .

Take care of your means, keep your Surplus on intrest and go right Straight ahead. . . .

W. H. Cook

JOSEPH PREPARES TO MARRY

Boston, June 22, 1877

My Dear Parents,

It is seventeen years since I first met the amazingly perfect lady to whom you were introduced last autumn at the Astor House in New York. There has been abundant opportunity for acquaintance. As there was no engagement until within the last two years, & as no correspondence passed between us while I was abroad, there was also abundant opportunity for us to forget each other. Providence has not allowed us to do so. The delicate health of Miss Hemingway is the only point as to which we have either of us had any great hesitation. On that subject, several physicians have given their opinions. One of the foremost specialists in New York, a writer of the very highest authority, was lately consulted, & he bids us go forward. There has been no step taken in the dark. . . .

In some way unknown to either of the parties concerned, the news of the prospective marriage has reached the Boston & New York newspapers, & has been copied everywhere. The ceremony will take place at New Haven, June 30. On that day send us your most fervent blessings.

Your affectionate son,

Joseph Cook

In 1876, Joseph and his parents visited the Philadelphia Centennial Exposition. On the way home they stopped at the Astor House where Joseph arranged a meeting between his father and mother and the future Mrs. Joseph Cook. It is this event to which Joseph refers in the preceding letter.

HIS FATHER'S COMMENTS

Ticonderoga June 27 / 77

My Dear Son,

I have thought mutch of the change of life you are about to make. I have been glad and Sorry that it was to be So, but uppon the whole I think its all for the best considering your age.

We Should be glad to be presant, but our age and distance are obstacles in the way. Excuse us & except our Prayers for your future welfare.

Your prospects look favorable in the future. Many envy your position who will place obstacles in your path but a man that is able to leed Boston can bag them all. . . .

Mutch Love for Georgia your companion. May the good Angels attend you in your Bridal tour and thro life. May She be able to add largely to your temporal and Spiritual interests. May you both become as beacon lights to lead the Lost to a Haven of rest. . . .

The knowing ones have goseped So mutch about your getting married that its become a thing of the past. The great interest with them now is to See the bride and then a little More talk and all is in the past.

Wm. H. Cook

LECTURES IN NEW YORK

Ticonderoga Oct. 1st 1877

My Dear Son,

Saturday evening Moses brought the big mail, the Book, the cloak for Kit, the picture from the Dutchess which is verry life like, the good letter from Joseph giveing better news than I would durst ask for, if a voice from Heaven had Said ask what thou will and it Shall be given. We Spread them all out on the table and thanked the Lord for his goodness. Your Salery is now Satisfactory. Away with your pop gun prices and take the lines of a double team.

Williams and Boston just begin to appreciate your worth. Study N york and try to do the wicked citty Some good. The Soil must be mellowed by the confession of Some of the ring. . . . Wall we have thought mutch of your first lecture, watching the time and ending of the hour. . . .

I like the getting up of your book. Its been put thro on a rush out in the right time. May it be carried to the people as bountifull as the Sun light. . . . Look well to your income. You ought to lay up ten thousand dollars the comeing year. . . .

Your father
Wm. H. Cook

Joseph's name for his wife was "Duchess"; hence the allusion in this letter.

JOSEPH'S BOOK A BEST SELLER

Boston, Oct. 8, 1877

My Dear Parents,

My book on Biology has gone through five Editions in six days. It was published Sep. 29; &, on Oct. 6, four thousand copies had been sold & seven thousand printed.

A cataract equinoctial rain was falling in Brooklyn Thursday night, but I had 1000 hearers, & the Tribune's notice of the lecture was favorable enough, as you saw.

The first audience in Tremont Temple was large & of the best quality. The course opened auspiciously. The second audience today was yet larger, & I feel quite sure that the Boston Lectures are starting thoroughly well.

Such announcements are being made in New York that I think I shall not start there with a stumble.

Your affectionate Son,

Joseph Cook

HIS AUDIENCES GROW

Boston, Oct. 15, 1877

My Dear Parents,

The Supreme Powers are kind to the Son you Educated.

It is estimated that some 100,000 copies of the Boston Monday Lectures are now published in newspapers every week.

I sold the copyright of the present course to Mr. Waters of the Advertiser for $1000.

He has sold the right of republishing to the New York Independent & the Cincinnati Gazette & to the New York Advocate.

The latter paper alone has 55,000 subscribers.

A weekly republication is announced in London.

Both the commendations & the criticisms in the Independent of last week help the sale of my book. Why the Independent gives so much space to the Boston Monday Lectureship, I do not know. There have been to my knowledge no requests made by my friends for attention from that paper.

Today the Tremont Temple audience was larger than heretofore this season, & more enthusiastic.

Your affectionate Son,

Joseph Cook

THE MONDAY LECTURER

"MY CUP RUNNETH OVER"

Ticonderoga Oct. 21st 1877

My Dear Son, & Daughter,

. . . It gives me great joy that your labours are partly in N york Citty. . . . Proffessor Peaboddy of Harvord, in the Independant places you Second to no man in the world. Providance keep you from Stumbling from this high Standard; I tremble when I look uppon your high Standing. . . .

A clergyman from port Henry that changed with Rev. Jones called a few days a go to See the parents of Joseph [who] as he Said was the comeing man of the times. . . . My highest ambition is fully Satisfied. Give god the glory. How richly I am paid for all my labour. . . .

The world is your field. No pent up city holds the[e]. Hip, Hip, Horrah good.

Your father
Wm. H. Cook

CHAPTER VIII

The Rising and the Setting Sun

IN 1880 Mr. Cook began to suffer from an ailment common to old age. Joseph sent his Boston physician to Ticonderoga, who prescribed a treatment that gave some relief. Joseph felt that he and his wife might safely embark upon a world tour lasting two years. They visited England, Scotland, and India, in which countries Joseph addressed tremendous audiences. He went on to Australia, where his success was equally great.

HEART'S DESIRE ATTAINED

Ticonderoga Jan 26th 1880

Forty two years ago, in 1838 a child was born at what is now called Cliff Seat of humble parents Strugling with poverty, oweing for a great part of their posesions, and now looking back forty two years over the History of this famely and child, considering the vast amount of Sunshine God has bestowed uppon this famely, especialy this child which was dedicated to the Lord on the night of his birth, we have great reason to rejoice and be exceeding glad for what the Lord has done for us.

This child has now becom a man that is a teror to evel doers, Speaking to Millions by his words and thro the Press. Mutch joy has been brought to this famely by his Success, by his freedom from bad habits, by his thorough going virtuous principles. The desire of my heart is completely filled.

I hope you will not over work. If the harness begins to chafe hold up. Health and constitution is worth more than money. . . .

I cannot bear the thoughts of your going abroad. We must not let Simpathy rule. Its no doubt but the time has come for you to cross the waters again. . . .

Your father
Wm. H. Cook

THE OLD HAND TREMBLES

Ticonderoga Feb 15th 1882

Dear Children,

Your Letter of the 12th Jan from Bombay came the 14th of Feb. Were verry glad to hear that you had So plesant quarters, and that your Lectures opened So well. . . . Well, there is thirty days between our knolledge of each other, whitch makes it Seem verry lonely. I think that you will be able to do good in India. . . .

My condition gives me a great deal of unrest. I don't know why I am so aflicted with pain. Almost two years Since the first Doctor was called. Can I stand it two years more? God only knows. . . .

Wm. H. Cook

Excuse my hand trembling.

TOWN NEWS

Ticonderoga March 30th 1882

My Dear Children,

We have just Recd G. Letter Saying that you Should be at home in Six months, and that you talked of changeing your business to the Readpath beauro. This Seames to bring you nearer to us. The women are just begining to clean house. Robt works for me for Six months. I have to pay him 20. Dollars pr month. I have hired a boy for Six months for 10 dollars pr month. So my help is engaged for the Summer. My health is about the same.

We are to have a bank built in town by the Orwell bank men. Gilegan and Stevens are putting up a 4 Story building. Drake is building a large building where the old brick Store used to Stand. . . .

Horace and Genna Moses are off to Po[u]ltney to School full of ambition. The Stalk have wintered quite well. . . . The Lord bless guide and keep you.

Wm. H. Cook

Horace Moses referred to in the preceding letter was Horace A. Moses, later founder and president of the Strathmore Paper Company, West Springfield, Massachusetts. Mr. Moses has been a generous benefactor of his native town. The hospital at Ticonderoga, the Liberty Monument in memory of the French, Indian, British, and Colonial soldiers who fought at Ticonderoga, the Community Building, and also the John Hancock House, a replica of John Hancock's house in Boston, given by Mr. Moses as headquarters of the New York State Historical Association, are among the examples of his munificence.

THE OLD FOLKS BUTCHER

Ticonderoga Dec 7, 83
1 oclock a.m. Friday

Dear Georgie

You are a good girl to write us so many letters and to give us so nice a description of your Thanksgiving visit. I can fancy what a pleasant time you must have had. We thought of you much, while we were eating our chicking pie, and other goodies, with no company but our hired help. We were all made quite sweet in the Evening by the arrival of your box of candy. We all partook of it even to Vim, and he smacked it down, just like any little dog. . . .

Well, we have been in quite a muss all this week. We killed our hog, and cow on Monday, and have made Sausages, head-cheese, tried lard, and tallow, and today I feel as if I had got through butchering, for I am all done except boiling my meat for mince pies. We wish you were here to help us eat up some of our goodies for the weather is so warm that we are afraid it will spoil. We have had lots of Porters house steak, besides some chickings. Pa and myself have worked very hard this week, and he is feeling pretty much down today, as the excitement is over. . . . But I am feeling first rate today, for I took some of quieting medicine last night, and had a good night's sleep. . . .

I hope we shall have Sleying soon now, for we have now got all ready for winter and I want it to come. I have not done much on my Afghan yet. Have not had much time. I have 2 stripes knit. I like it very much. The Cloth is good, but my colers are not quite as good as I should like. I could not get all the shades I wanted, so I took some that was left of yours and coulered it, and had very good luck. . . .

How I wish I could be with you. But no—all such pleasures are not for me. Home is my place, and I must be content. Sometimes I feel gloomy, and think I am of no consequence, only to work, but I drive it of and say all right, as long as I am able to do, so long I must be content and try to do for others. . . .

This from your affectionate
Mother Cook

"A WARM BLANKET OVER ME"

Ticonderoga Jan 24th 1884

Dear Son,

Forty six years ago, come the Jan. 26, you were born a 3½ lbs. baby. Providance has guided you for 46 years in a most wonderful way. Your life has been preserved amid many dangers, uppon land and Sea. Places have been opened for you to do good and you have filled those places to the glory of god. Your Success has been a warm blanket over me, especially Since the Spring of 1880, when I was brought under the cloud of lothesom difficulty and pain.

The winter is verry cold. Cellar freezes. Good Sleighing. Stalk doing well, all things concidered. Breaking colts a little. I do not do mutch but keep fires. . . . God bless you.

W. H. Cook

A FEW LAST WORDS OF WISDOM

March 21St 1884

Dear Son,

I have just been reading the coments on your last Lecture. I think that the papers will be full of criticism as you leave for your Lecture tour. Let the press do their own fiting. If your efforts are of the Lord they will live. Try to be cortious and avoid personal combat. The Lord help you to be wise as a Serpant and harmless as a dove. Take care of yourself. You are a prominent target for evil doers. Hit the enemy hard blows uppon the platform, and avoid personal attacks. All well at home. Thanks for your good letter. We dont want for any thing. The snow is Slowley disapearing. Spring will Soon be uppon us. . . . The blessings of God go with you and protect you in the future as he has in the past. . . .

Wm. H. Cook

AND SO FAREWELL

Ticonderoga Jan 23d 1885

Dear children,

Thermomator 20 below zero. Evrything froze up. Verry hard for the Stalk. Lambs dieing. All well, 12 hands on the house working, Some painting, Some puttying in glass, Some putting ceiling overhead on piasies. Its verry cold work. Goes Slow. The kitchens nearly done. Three Stoves are running besides the oil Stove. The blinds will Soon be here.

I have read Ben Hur, a thrilling novel. . . . Your House is a beauty. . . . Good Sleighing. George driving colts. May the good Spirit guide you.

Wm. H. Cook

This was his last letter to his son. The day soon came when he was obliged to take to his bed. The flocks and horses knew their master no more. Joseph was hastily summoned from the lecture platform. He arrived to find his father sinking. They exchanged mutual protestations of their faith until the morning of the nineteenth of March, 1885, when the spirit of William H. Cook took flight from Trout Brook Valley. His wife Merrette survived him five years. They lie side by side in a little cemetery not far from the farm where they spent their lives.

* * * * *

The new house that was a "beauty," as Farmer Cook described it in his last letter to his son, was a large addition to the north side of the parental abode, which Joseph constructed when fame and fortune grew. Its distinguishing architectural features were a broad veranda commanding a view of the valley, bay windows that were popular in those days, and two towers rising above the roof, one designed as a study for Joseph Cook and the other for his wife. This was their summer home, and there they invited kindred souls in the fields of religion and reform. Among the most notable of the visitors who often enjoyed their hospitality were Emma Willard and Anthony Comstock.

Joseph Cook continued to flourish as a lecturer for many years after his parents' death. In 1895 he went for the second time to Australia. There twenty years of public speaking, with long periods of constant travel by day and oratory by night, made themselves felt even on Joseph Cook's large body. He was stricken with nervous prostration, and journeyed from Australia to Japan where Mrs. Cook met him. They returned to the United States in December. He recuperated at Clifton Springs, New York, and thereafter at Cliff Seat. Although at length he recovered his health sufficiently to speak again in public, and to resume writing for his favorite publications, his former vigor never returned. He and Mrs. Cook continued to spend their summers at Ticonderoga and made their winter home at Newton Center, Massachusetts. In the winter of 1900, he delivered four of the series of Boston Monday Noon Lectures, thus bringing the total of these discourses to two hundred and fifty since 1875. He died at Cliff Seat June 24, 1901. His widow continued to reside there until she died August 3, 1921. Their marriage had been childless.

Before many years Cliff Seat fell into decay. Nothing but dilapidation remains to tell the story of simple life and country ways, of hopes and fears, ambition, perseverance, and triumph that had been unfolded there.

LaVergne, TN USA
20 January 2010
170696LV00001B/59/P